AF566652

BY-PRODUCTS AND WASTE MATERIALS IN FAT TECHNOLOGY

ELLIS HORWOOD SERIES IN APPLIED SCIENCE AND INDUSTRIAL TECHNOLOGY

Series Editor: Dr D. H. SHARP, OBE, former General Secretary, Society of Chemical Industry; formerly General Secretary, Institution of Chemical Engineers; and former Technical Director, Confederation of British Industry.

This collection of books is designed to meet the needs of technologists already working in the fields to be covered, and those new to the industries concerned. The series comprises valuable works of reference for scientists and engineers in many fields, with special usefulness to technologists and enterpreneurs in developing countries.

Students of chemical engineering, industrial and applied chemistry, and related fields, will also find these books of great use, with their emphasis on the practical technology as well as theory. The authors are highly qualified chemical engineers and industrial chemists with extensive experience, who write with the authority gained from their years in industry.

Published and in active publication

PRACTICAL USES OF DIAMOND
A. BAKON, Research Centre of Geological Technique, Warsaw, and A. SZYMANSKI, Institute of Electronic Materials Technology, Warsaw
RECOVERY AND REHABILITATION OF CONTAMINATED LAND
R.J.F. BEWLEY, Biotreatment Limited, Cardiff, and D.H. SHARP, O.B.E., Sevenoaks, Kent
NATURAL GLASSES
V. BOUSKA *et al.*, Czechoslovak Society for Mineralogy & Geology, Czechoslovakia
INTRODUCTION TO PERFUMERY: Technology and Marketing
TONY CURTIS, Senior Lecturer in Business Policy and International Business, Plymouth Business School, and DAVID G. WILLIAMS, Independent Consultant, Perfumery and Director of Studies, Perfumery Education Centre
POTTERY SCIENCE: Materials, Processes and Products
A. DINSDALE, lately Director of Research, British Ceramic Research Association
THE HOSPITAL LABORATORY: Strategy Equipment, Management and Economics
T.B. HALES, Arrowe Park Hospital, Upton, Wirral
ASPECTS OF PRESSURE CONTROL AND SAFETY IN OFFSHORE DRILLING OPERATIONS
R. HOSIE, formerly Deputy Head, Mechanical and Offshore Engineering, Robert Gordon's Institute of Technology, now of Hosie Engineering Limited, Aberdeen
OFFSHORE PETROLEUM TECHNOLOGY AND DRILLING EQUIPMENT
R. HOSIE, formerly of Robert Gordon's Institute of Technology, Aberdeen
MEASURING COLOUR: Second Edition
R.W.G. HUNT, Visiting Professor, The City University, London
CHARACTERIZATION OF FOSSIL FUEL LIQUIDS
D.W. JONES, University of Bristol
LABORATORIES: Design, Safety and Project Management
Editor: T.J. KOMOLY, ICI Engineering, Winnington, Ches.
PAINT AND SURFACE COATINGS: Theory and Practice
Editor: R. LAMBOURNE, Technical Manager, INDCOLLAG (Industrial Colloid Advisory Group), Department of Physical Chemistry, University of Bristol
ECONOMICS AND FINANCIAL STUDIES FOR ENGINEERS
D.J. LEECH, Senior Lecturer, Department of Management Science, University College of Swansea, Wales
CROP PROTECTION CHEMICALS
B.G. LEVER, International Research and Development Planning Manager, ICI Agrochemicals
HEALTH PROTECTION FROM CHEMICALS IN THE WORKPLACE
Editor: Dr P. LEWIS, Senior Executive, Occupational Health, Chemical Industries Association, London
HANDBOOK OF MATERIALS HANDLING
Translated by R.G.T. LINDKVIST, MTG, Translation Editor: R. ROBINSON, Editor, *Materials Handling News.* Technical Editor: G. LUNDESJO, Rolatruc Limited
FERTILIZER TECHNOLOGY
G.C. LOWRISON, Consultant, Bradford
NON-WOVEN BONDED FABRICS
Editor: J. LUNENSCHLOSS, Institute of Textile Technology of the Rhenish-Westphalian Technical University, Aachen, and W. ALBRECHT, Wuppertal
REPROCESSING OF TYRES AND RUBBER WASTES: Recycling from the Rubber Products Industry
V.M. MAKAROV, Head of General Chemical Engineering, Labour Protection, and Nature Conservation Department, Yaroslav Polytechnic Institute, USSR, and VALERIJ F. DROZDOVSKI, Head of the Rubber Reclaiming Laboratory, Research Institute of the Tyre Industry, Moscow, USSR
ACCIDENTAL EXPLOSIONS: Volume 1: Physical and Chemical Properties
LOUIS A. MEDARD, formerly Head of the Laboratory in the French Government Explosive Branch
ACCIDENTAL EXPLOSIONS: Volume 2: Types of Explosive Substances
LOUIS A. MEDARD, formerly Head of the Laboratory in the French Government Explosive Branch

series continued at back of book

BY-PRODUCTS AND WASTE MATERIALS IN FAT TECHNOLOGY

HENRYK NIEWIADOMSKI (deceased)
formerly of the Technical University of Gdańsk, Poland

HANNA SZCZEPAŃSKA (deceased)
formerly of the Institue of Industrial Chemistry, Warsaw, Poland

Translator Editior
K. G. BERGER, London, UK

ELLIS HORWOOD
NEW YORK LONDON TORONTO SYDNEY TOKYO SINGAPORE

PWN – POLISH SCIENTIFIC PUBLISHERS
WARSZAWA

English edition first published in 1995
in coedition between
ELLIS HORWOOD LIMITED
Market Cross House, Cooper Street,
Chichester, West Sussex, PO19 1EB, England

A division of
Simon & Schuster International Group
A Paramount Communications Company

and

POLISH SCIENTIFIC PUBLISHERS PWN Ltd.
Warsaw, Poland

Revised and enlarged translation from the Polish edition
Produkty uboczne i odpady tłuszczowe. Wykorzystanie i wpływ na środowisko, Państwowe Wydawnictwo Naukowe, Warszawa 1989

Translated by Tadeusz Górecki

Printed in Poland
ISBN 83-01-11652-8 (PWN)

British Library Cataloguing in Publication Data

A catalogue record for this book is available from the British Library

ISBN 0-13-109547-1 (Ellis Horwood)

Library of Congress Cataloging-in-Publication Data

Available from the Publishers

Table of Contents

SYMBOLS AND ABBREVIATIONS 9

INTRODUCTION *H. Niewiadomski* 11

Chapter 1 ORIGIN OF THE BY-PRODUCTS AND WASTES . . 13
- 1.1 Edible oil industry *H. Niewiadomski* 13
  - 1.1.1 Seed pretreatment 13
    - 1.1.1.1 Seed cleaning 14
    - 1.1.1.2 Seed drying 14
    - 1.1.1.3 Dehulling 15
  - 1.1.2 Processing of seed 18
    - 1.1.2.1 Pressing 19
    - 1.1.2.2 Solvent extraction 20
    - 1.1.2.3 Solvent losses 22
    - 1.1.2.4 Application of supercritical fluid 27
    - 1.1.2.5 Other methods of the extraction of fats 29
  - 1.1.3 Processing of palm fruit 31
    - 1.1.3.1 Sterilization of fruit 31
    - 1.1.3.2 Stripping of fruit 32
    - 1.1.3.3 Digester and screw press 33
    - 1.1.3.4 Settling and clarification of oil 33
    - 1.1.3.5 Recovery of palm kernels 33
    - 1.1.3.6 Yield of oil mill products 34
  - 1.1.4 Processing of olives 34
  - 1.1.5 Degumming of oils 35
    - 1.1.5.1 Degumming of soyabean oil 36
    - 1.1.5.2 Degumming of rapeseed oil 39
    - 1.1.5.3 Desludging 40
  - 1.1.6 Obtaining of lecithin 43
  - 1.1.7 Refining of oils 46
    - 1.1.7.1 Removal of suspensions 46
    - 1.1.7.2 Dewaxing 48
    - 1.1.7.3 Alkali neutralization of oils 51

1.1.7.4 Bleaching of oils 55
1.1.7.5 Deodorization of oils 59
1.1.7.6 Losses during alkali refining 66
1.1.7.7 Physical neutralization 67
1.1.7.8 Neutralization in miscella 70
1.1.7.9 Winterizing of oils 70
1.1.8 Hydrogenation of oils 70
1.2 Non-edible fat processing industry *H. Szczepańska* 71
1.2.1 Refining of non-edible fats 73
1.2.1.1 Methods of animal fats refining for various applications 74
1.2.1.2 Methods of refining of fats for the production of fatty acids and glycerine 75
1.2.2 Hydrolysis of fats 79
1.2.3 Fatty acids distillation 84
1.2.4 Fractionation of fatty acids 88
1.2.4.1 Fractional distillation 89
1.2.4.2 Crystallization 93
1.2.4.3 Fractionation by hydrophilization 95
1.2.5 Hydrogenation of fatty acids 96
1.2.6 Obtaining of fatty alcohols 99
1.2.7 Saponification of fats and fatty acids 102
1.2.8 Processing of specific non-edible fats 102
1.2.8.1 Processing of wool wax 102
1.2.8.2 Castor oil processing 105
1.2.8.3 Processing of tall oil 107
1.2.9 Processing of other fatty derivatives 111
1.2.9.1 Fatty acid methyl esters 111
1.2.9.2 Fatty acid monoglycerides 112
1.2.9.3 Oxyethylated derivatives 115
1.2.9.4 Sulphated and sulphonated fatty alcohols and acids 115
1.2.9.5 Alkanolamides 116
1.2.9.6 Fat amides 117
1.2.9.7 Fatty nitriles 117
1.2.9.8 Fatty amines 118
1.2.9.9 Acid or ester dimers 118
1.2.9.10 Polymerized oils 119
1.2.9.11 Epoxidized oils 120
1.3 Other industries *H. Szczepańska* 120
1.3.1 Meat processing 120
1.3.2 Feed processing 121
1.3.3 Marine fat processing *H. Niewiadomski* 123
1.3.4 Tanning 124
1.3.5 Bone processing 125
1.3.6 Poultry processing 126
1.3.7 Others 126

1.3.7.1 Breeding of furred animals 126
1.3.7.2 Sewer grease 126
References 128

Chapter 2 COMPOSITION, PROPERTIES AND UTILIZATION OF BY-PRODUCTS 134
2.1 Oil industry *H. Niewiadomski* 134
2.1.1 Meal 134
2.1.2 Lecithin 138
2.1.2.1 Composition and properties 138
2.1.2.2 Utilization of lecithin 142
2.1.3 Soapstock and post-refining acids 144
2.1.3.1 Composition and properties 144
2.1.3.2 Utilization of soapstock 145
2.1.4 Fats from adsorbents used for bleaching 147
2.1.4.1 Composition and properties 147
2.1.4.2 Utilization of spent bleaching earth 147
2.1.5 Deodorizer scum 148
2.1.5.1 Composition and properties 148
2.1.5.2 Utilization of deodorizer scum 149
2.2 Non-edible fat processing industry *H. Szczepańska* 150
2.2.1 Glycerine 150
2.2.1.1 Sweet waters 150
2.2.1.2 Soap lyes 152
2.2.1.3 Other by-products containing glycerine 153
2.2.2 Distillation residue 154
2.2.2.1 Composition and properties 154
2.2.2.2 Utilization of distillation residue 156
2.2.3 Forerunnings 157
2.2.4 Monomers 158
2.2.5 Lanolin soaps 158
2.3 Other industries *H. Szczepańska* 159
2.3.1 Slaughter and poultry fats 159
2.3.2 Fish oils 160
2.3.3 Utilization grease, degras, bone fat 161
2.3.4 Tall oil 163
2.3.5 Suints 164
2.3.6 Furred animal fats 164
References 164

Chapter 3 COMPOSITION, PROPERTIES AND UTILIZATION OF WASTES *H. Niewiadomski* 167
3.1 Seed impurities 167
3.2 Hulls 168
3.3 Palm oil mill waste 169
3.4 Spent hydrogenation catalyst 169

3.5 Spent bleaching earth 170
3.6 Frying oils 170
3.7 Sediments and liquid wastes 170
3.8 Wastes from oleochemicals production 172
3.9 Wastes from fishing industry 173
3.10 Other wastes *H. Szczepańska* 173
3.10.1 Glycerine distillation residue 173
3.10.2 Wastes formed during the production of sebacic acid from castor oil 174
3.10.3 Wastes from meat processing plants amd mass nutrition utilities 174
3.10.4 Miscellaneous *H. Niewiadomski* 175
References 175

Chapter 4 PROTECTION OF THE NATURAL ENVIRONMENT *H. Niewiadomski* 177
4.1 Fumes 177
4.1.1 Dust 178
4.1.2 Fumes from solvent extraction 179
4.1.3 Fumes from fat refining 180
4.1.4 Fumes from fat rendering plants 181
4.2 Liquid wastes 182
4.2.1 Origin of liquid wastes 182
4.2.2 Waste treatment 185
4.2.2.1 Closed systems 189
4.2.2.2 Mechanical cleaning 194
4.2.2.3 Chemical treatment 196
4.2.2.4 Physicochemical treatment 197
4.2.2.5 Biological treatment 201
4.2.2.6 Biological treatment of palm oil mill effluent . 206
4.2.2.7 Biological treatment of olive oil mill effluent . 207
4.2.2.8 Treatment of wastes from oleochemical plants . 210
4.2.2.9 Profitability of waste treatment 212
4.2.2.10 Standards of wastes purity 213
4.3 Solid wastes 213
References 215

INDEX 217

Symbols and Abbreviations

AnV — anisidine value
AOM — active oxygen method
AV — acid value
BOD — biological oxygen demand
b.p. — boiling point
COD — chemical oxygen demand
c.p. — congeal point (titer)
CTO — crude tall oil
DFA — distilled fatty acids
d.m. — dry matter
DTO — distilled tall oil
d.w. — dry weight
FA — fatty acids
FFA — free fatty acids
GLC — gas-liquid chromatography
HPL — hydratable phospholipids
i.p. — ignition point
IV — iodine value
m.p. — melting point
NPL — non-hydratable phospholipids
O — rapeseed variety with erucic acid content of oil less than 5%
OO — rapeseed variety with erucic acid content of oil less than 5% and the glucosinolate content in seed below 3 mg/g
p — pressure
PL — phospholipids
PV — peroxide value
SFA — fatty acids from splitting (crude FA)
SM — saponifiable matter
SV — saponification value
TFA — total fatty acids
TLO — tall light oil
TOFA — tall oil fatty acids
TOP — tall oil pitch
TOR — tall oil rosin
TP — theoretical plate
UM — unsaponifiable matter

Introduction

In the edible and non-edible fats industries, apart from the major products, by-products are also formed, whose quality and quantity are more or less constant. These by-products are often badly utilized. Proper utilization would increase the profitability of fat processing plants. Lecithin and post-refining acids belong to such products in the oil industry, while in the non-edible fats industry distilation residues and sludges.

As in all industry, in fat processing plants waste materials are also formed. Whether the materials formed during raw materials processing are considered as by-products or wastes is highly arbitrary.

In order to simplify the definition it can be assumed that all the materials, irrespective of their physical state, leaving the processing line other then the main or by-products, can be regarded as wastes. In the case of the oil industry these will be waste materials formed during seed pretreatment, extraction of oil, refining, hydrogenation and production of final edible products. These waste materials can occur in a liquid form, e.g. in liquid wastes, in a solid form, e.g. deoiled bleaching earth or spent catalyst, as well as in a gaseous form, e.g. fumes formed during deodorization or dusty vapours from the meal desolventizer.

Rationalization of by-product and waste management is one of the main research problems of the industry. It is also closely related to the environmental regulations, which are becoming increasingly rigorous. The basic trend is to enable the regeneration or other utilization of wastes, the final goal being their disposal in a manner that does not interfere with the protection of the environment.

The scope for technical utilization of wastes and the level of environment protection depend on the technological level of a given country. Processing of particular wastes aimed at obtaining usable products is not profitable everywhere. Also the industry does not always want to incur the capital and operating costs connected with protecting the environment against the destructive impact of production. It is necessary therefore to develop better ways of utilization of fatty by-products and of fatty wastes both from the fat industry, i.e. from oil manufacturing and non-edible fats processing, and from the related industries.

The amount and qualitative characteristics of by-products and wastes depend to a certain extent on the kind of the main technological process and the equipment used. Technological progress aims at a development of technologies yielding little or no wastes (Kubicki, 1983). In the fat industry this is accomplished by more accurate separation of fats from other components of the raw materials, e.g. proteins or carbohydrates. Hence, in considering the technological systems in which by-products or wastes are formed, it is purposeful to discuss not only traditional processes, but also recently developed ideas, since it is possible that they will be implemented in the future.

CHAPTER 1

Origin of the By-Products and Wastes

Fatty wastes are also formed in industries other than the fat industry. In the processing of non-edible fats the basic raw materials are waste animal fats from meat and utilization plants, post-refining acids, bone and tannery fats, fish oils, etc. They are therefore by-products and wastes formed most often in non-fat technologies. On the other hand, in the fat industry not only fatty, but also other wastes are formed, due for example to the presence of non-fatty components of the raw material (e.g. contaminants of oil seeds) or to the technological processes (e.g. catalyst). This book deals therefore not only with fatty wastes, but also with others connected with the fat industry.

1.1 EDIBLE OIL INDUSTRY

Research aiming at decreasing the costs of oil production concerns almost exclusively the main products, i.e. edible oils and margarine. On the other hand, there is little research in recent years concerning a better utilization of by-products and wastes.

By-products are in this case mainly protein substances separated from fatty raw materials in the form of meal or oil cake, phospholipids commonly called lecithin, post-refining acids, fats contained in the adsorbent used for bleaching and the deodorizer scum. Wastes are formed during seed pretreatment, extraction of fats, degumming, neutralization, bleaching, deodorization, oil polishing and hydrogenation, as well as production of final products. This classical outline of operations and processes is being changed during recent years by the combination of certain processes or operations (e.g. neutralization with deodorization). Technological progress influences therefore the amount and quality of the by-products and wastes formed.

1.1.1 Seed pretreatment

Vegetable oil raw material delivered to a fat processing plant is cleaned, usually dried and sometimes dehulled. All these operations yield wastes that can partly be used as by-products.

1.1.1.1 *Seed cleaning*

Seed delivered to an oil mill always contains impurities, which must be removed before or after drying. These impurities are: fragments of stems, leaves, other seeds, sand and other impurities. The amount of these impurities depends on the accuracy of work of the seed deliverer. Impurities from field and transport deteriorate the quality of the extracted fat and the meal or oil cake produced, as well as damage the machines. The outline of the traditional manner of seed cleaning is presented in Fig. 1.1. The separated waste material consists of stones, metals, sand and soil, as well as light impurities like hulls and dust.

An important reason for purifying the seed is the dust explosion hazard. This hazard arises during transport of seed by conveyors and during storage in elevators. Explosions in stores of oil mills occur not infrequently, and though their probability depends on a number of parameters, yet the presence of dust is the most important factor inducing this phenomenon (Zockoll, 1982).

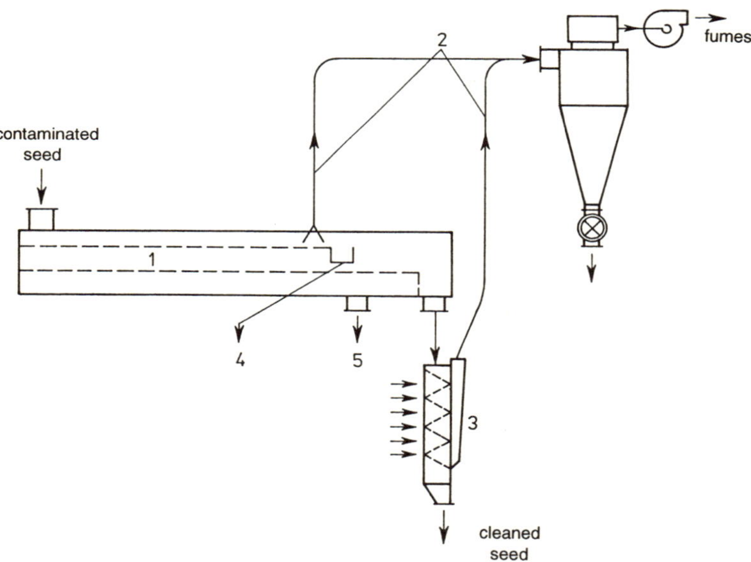

Fig.1.1 Traditional manner of seed cleaning (Moore, 1983): 1—screen, 2—outlet of light impurities, 3—aspirator, 4—stones and metal pieces outlet, 5—sand outlet.

1.1.1.2 *Seed drying*

Drying is necessary for seed whose humidity exceeds the level allowable during storage. This operation is accomplished by a removal of humid air carrying away light impurities, which are captured in cyclones.

1.1.1.3 Dehulling

Certain seeds (e.g. sunflower) require dehulling. This operation is recently carried out more frequently, since it affects the quality of the extracted meal and oil positively and increases the oil yield.

Removal of soyabean hull. To facilitate this operation the soyabean is dried to 9.5–10 % water content after storage (Moore, 1983). The further steps of the preliminary operation are presented in Fig. 1.2. The seeds are crushed to 1/4–1/6 of their original size.

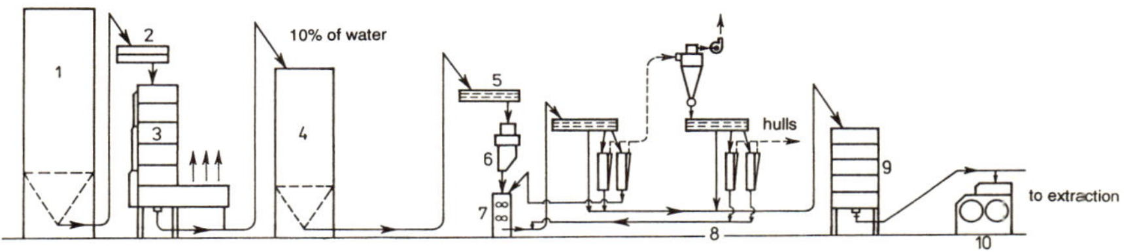

Fig. 1.2 Pretreatment of soyabean (Moore, 1983): 1—seed store, 2—cleaner, 3—drier, 4—apparatus for tempering the seed for 1–3 days, 5—cleaner, 6—scales, 7—crushing mill, 8—huller, 9—cooker for conditioning, 10—flaker.

After passing the seeds through a cracking mill the hulls are separated from meats and subjected to aspiration, while the uncrushed seeds are recirculated and passed again through the rollers. The liberated kernels pass onto a second sieve of smaller mesh, on which small impurities are sifted. At the end of this sieve the remaining hulls are again aspirated. Tiny impurities are combined with the main seed supply. The separated hulls contain a certain amount of kernels. Lighter hulls are separated from heavier kernels by aspiration, which minimizes losses. Hence, in the system described, wastes in the form of hulls, drier fumes from aspiration and volatile dust from dehulling are formed. Depending on the separation accuracy required an intermediate fraction can also be received and subjected to an additional separation (Swern, 1964). Another method is the separation in brine, utilizing the difference in the density of hulls and kernels. This separation is very accurate, but the kernels require drying before further processing.

The composition of wastes formed in the Escher–Wyss process, introduced to the industry a few years ago, is slightly different. This system utilizes fluidization for drying, dehulling and conditioning of the soyabean. In order to decrease the consumption of the energy of flow, the air is recirculated between the particular elements of the equipment (Fig. 1.3). The seed taken from the store is introduced after cleaning to a fluidized-bed drier, where its water content is decreased by 1–3%. It is subsequently passed through a set of rollers, where it is divided, and reaches an impact mill, where the halves of seeds are separated from the hulls. The bulk passes next through a multi-stage aspirator supplied with heated air sweeping away the hulls.

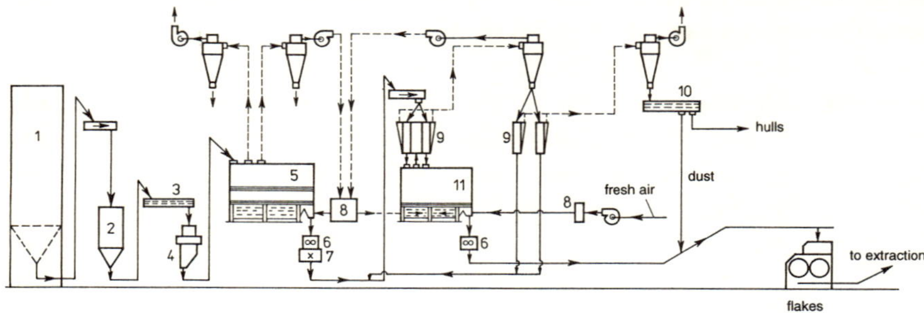

Fig. 1.3 Use of fluidization for seed pretreatment (Moore,1983): 1—seed store, 2—diurnal silo, 3—cleaner, 4—scales, 5—drier, 6—crusher, 7—grinding mill, 8—cooker, 9— huller, 10—screen, 11—conditioner.

In this system the amount of air leaving the equipment through cyclones is greater, but the tiny seed particles swept away by its stream are recycled to the huller, which decreases the losses (Moore, 1983). The hull toaster used so far, consisting of a set of plates heated with steam, is replaced more and more frequently with a vertical apparatus, in which steam supplied directly to the bottom moisturizes the hulls when passing through thick layers of them. The conditioned hulls are subsequently passed through a section supplied with ambient air, decreasing the temperature and removing the excess

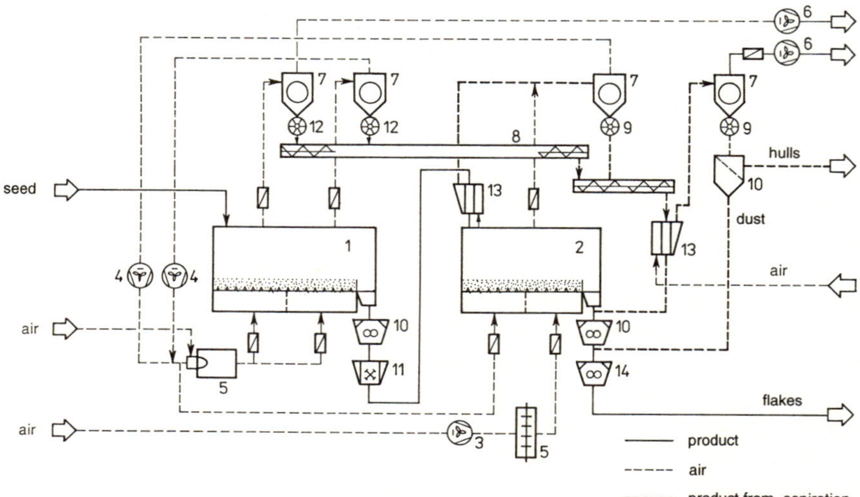

Fig. 1.4 Fluidized dehulling (Florin and Beretsch, 1983): 1—fluidal heater, 2—conditioner, 3—fresh air blower, 4—blower for air recirculation, 5—air heater, 6—exhaust fan, 7—cyclone, 8—helical conveyor, 9—rotating air lock, 10—crusher, 11—beater mill, 12—huller, 13—aspirator, 14—flaker.

water. Hulls prepared in such a manner are ground and stored. Another system of fluidized-bed dehulling is presented in Fig. 1.4. Soyabean seeds are heated in an apparatus to 75–92°C (at the seed surface). Heating causes opening and breaking of hulls detached from seeds. The contents are subsequently passed through a set of rollers and a disintegrator, where the seeds are divided and liberated from the hulls. Next the seeds are supplied through an aspirator to a fluidized bed conditioner, in which seed halves are conditioned by a stream of hot air in order to optimize the extraction. The product swept away by the aspirator is caught in a cyclone and recycled to the conditioner. The particles contained in the air from the aspirator are separated in a cyclone and next divided on a vibrating screen into hulls and dust, which is recycled to the main stream before the flaker (Florian and Bartesch, 1983). A comparison of the traditional dehulling system with the fluidized-bed one is shown in Fig. 1.5.

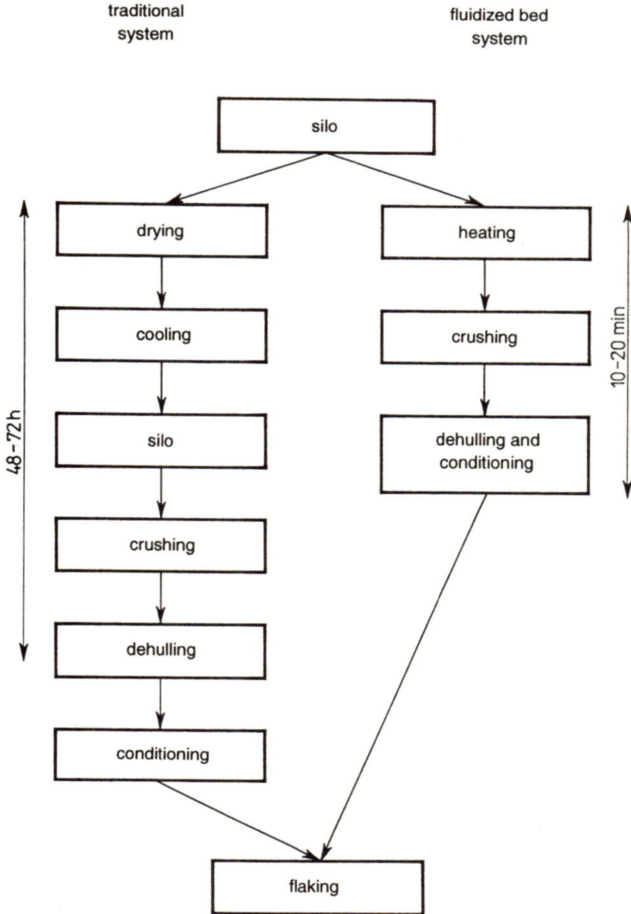

Fig. 1.5 Comparison of traditional and fluidized dehulling (Florin and Bartesch, 1983).

Depending on the protein and cellulose content of the soyabean, up to 25% of meal of 49% protein content is obtained. The remaining 75% of the obtained meal contains 44% of protein. When the separation of hull is accurate, the amount of the 49% fraction can increase even up to 40% (Fetzer, 1983).

Removal of sunflower hull. The fat content in hull is 1.8–2.8%. After a single dehulling the meal contains 35% protein, while after a repeated process this content increases up to 42%. From 100 kg of seed containing 28% of hull ca. 18.2 kg of hull can be removed, and the remaining 81.8 kg contains 72 kg of kernels and 9.8 kg of hull (Ruckenstein, 1982). Whereas conventional dehulling leaves 10–12% of hulls with the kernels, an airjet impact huller reduces the residual hull level to 2–3% (Tranchino *et al.*, 1984). The huller operates by sucking the seed into an air stream in a Venturi tube, accelerating the stream, and allowing the seed to strike an angled metal target. The hulls are separated using a combination of an air classifier and a fluidized vibrating table.

Removal of rapeseed hull. In order to yield the highest weight gain in feeding of swine and poultry the feed should contain little cellulose and more protein. The composition of fat contained in rapeseed hull is different from that of seed-leaf fat, since it contains ca. 10% less triglycerides, ca. 10% more polar lipids, and ca. 2% mono- and diglycerides. The AV is much higher, and the stability lower. These negative properties make this fat useless for feeding purposes, and indicate the need for rapeseed dehulling (Kozłowska *et al.*, 1988). Grinding of the meal and separation by aspiration does not yield good results (Roden, 1983). In 1985 dehulling of a double zero rapeseed was introduced on an industrial scale in one of the French factories (Chone, 1985). Decortication of rapeseed yields a meal with higher protein content and lower crude cellulose (Signoret, 1988). The seed is allowed to fall on a disc rotating at high speed and is flung against a ring-shaped metal target where the seed coat shatters. Separation of the seed coat, the kernels and the unbroken seed is achieved in a fluidized bed. The unbroken seed is recycled.

1.1.2 Processing of seed

Extraction of fats is usually accomplished by pressing in an expeller, coupling of pressing with solvent extraction, or solvent extraction only. Choice of one of these methods is made on the basis of economical calculations taking into account mainly the processing costs and the value of the products obtained, i.e. fat and the residue (meal or oil cake).

1.1.2.1 Pressing

Pressing is applied increasingly often. Use of this method avoids the escalating costs of solvent and protects the environment against fumes containing hydrocarbon. Changes in this direction have become possible owing to the application of expellers leaving only 3.5–4.5% of fat in the oil cake. A new type of preliminary press has been recently introduced to the industry. It accomplishes the functions of flaking, conditioning and prepressing. This press repeatedly crushes, cuts, squeezes and loosens the processed raw material. Mechanical energy used for this purpose is partly converted into heat, which increases the temperature. On the other hand, this type of press yields quite a lot of dust pressed together with the oil. This dust must be recycled, which decreases the capacity of the press, or has to be processed in an additional press. The advantages of the method are lower energy consumption and better quality of oil, containing less FFA, sulphur, chlorophyll and phosphorus.

Sunflower seed is pressed either after dehulling, or without this operation. Both methods have their advantages and drawbacks. Suspensions contained in the pressed oil should be accurately removed before storing, since they cause excessive losses during refining. The simplest method consists in passing the oil through a settling tank, in which the majority of the sediment settles down at the bottom. The sediment is collected and pumped to a sedimentation tank, from which it is gradually supplied to the ground seed at the inlet of the expeller. A more rational solution consists in an application of an automatic screen, in which oil flows through a sieve of 0.14–0.25 mm mesh. Impurities caught on the sieve move towards the end and fall to the outlet. After drying they are mixed with the ground seed. Pressed oil purified in this way is filtered through plate-and-frame filter presses. Filter cakes are mixed with the raw material directed to the expeller.

Another method of purification of the pressed oil, used in the case of rapeseed oil, is gravitational clarification in a continuous settling tank. The separated sediment is recycled through the toaster to pressing or is processed in a special press. Oil cake from this press is directed to solvent extraction together with the main stream of the raw material. The oil obtained from the press is recycled to the sieve settling tank in order to settle again the fine particles. The sediment can be directly subjected to solvent extraction together with the main stream, provided it does not hinder the process. Clarification of the pressed oil is often accomplished by means of centrifuges or filter paper filters, both leaving less than 0.02% suspension and requiring very little labour when automated. Pressed rapeseed oil contains 3–5% or 6–10% v/v of suspended matter. Double zero rapeseed yields more impurities than high erucic acid varieties, since the seeds are softer. Usually 65–75% of the suspended impurities are removed by sieves, while the rest is removed in settling tanks or by clarification.

In large plants special presses are used for processing these impurities. The oil cake from such presses can be easily solvent extracted. In cases when there is no such device the sediments are transferred directly to the expellers.

1.1.2.2 Solvent extraction

Solvent extraction with prepressing. Oil and meal are obtained from the solvent extraction. The value of the meal can exceed that of the oil, particularly when oil content in the raw material is low. In most cases, however, the meal is a by-product. Solvent extraction of oil seeds aims at decreasing the content of fat in the meal below 1%. Such results are achieved in the case of soyabean processing, while in the case of rapeseed processing they are difficult to achieve, probably due to difficulties in obtaining proper flakes (Beach, 1983).

An improved method of conditioning soyabean flakes prior to the extraction, called Alcon process, comprises the following operations between flaking and solvent extraction (Penk, 1981 a), (Fig.1.6):

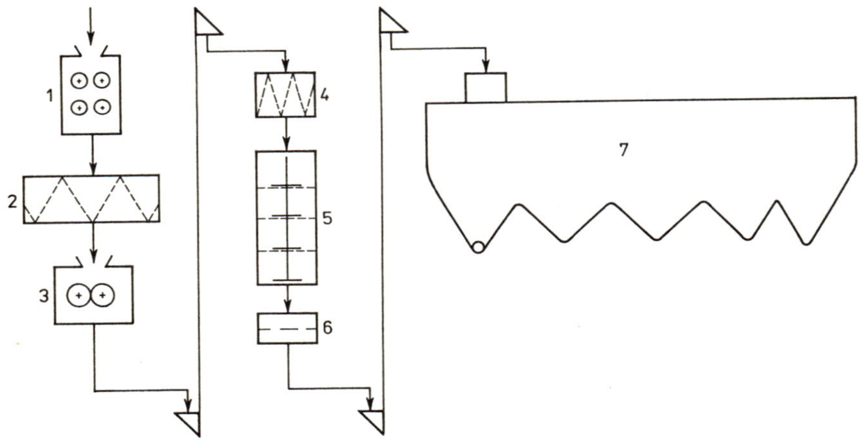

Fig. 1.6 Preparation of soyabean for extraction by the Alcon method (Penk, 1981a): 1—corrugated rollers, 2—seed treater, 3—plain rollers, 4—conditioner, 5—tempering equipment, 6—drying and cooling unit, 7—extractor.

- —conditioning by moisturizing with steam to 15–20% water content at a temperature of 95–110°C,
- —tempering for 15 min with continuous stirring, which causes agglomeration,
- —drying and cooling of flakes according to the requirements of the solvent extraction.

Flakes prepared in such a manner become granulated and reveal many positive properties during solvent extraction, e.g. better percolation, obtaining a meal of lower solvent absorptivity and higher quality (Penk, 1981 b), higher lecithin yield with greater hydration capacity, as well as a lower PL content of the crude oil. The Alcon system also causes an increase in the bulk density from ca. 0.36 kg/dm^3 to ca. 0.56 kg/dm^3, hence by 50%, which increases the capacity of the equipment by ca. 35%.

Modern extractors are operated under a slightly negative pressure, therefore they themselves are not the source of fumes. However, additional equipment, like water and solvent separators, terminal vent of the entire installation, foam trap, steam ejectors and cooling towers contribute to the composition and amount of the liquid wastes, emit fumes and heat and are sources of noise.

When the raw material is rich in fat, solvent extraction is usually used together with prepressing. The meal is subjected to a few operations aiming at acquiring the final quality appropriate for the best method of utilization, most often as a feed for slaughter animals. The meal leaving the extractor must be solvent-free, dried, cooled, sieved and sometimes pelletized. The first three operations are usually carried out in a desolventizer-toaster. In this device, apart from removing the solvent from the meal (initial content 25–40%), its water content is also reduced. Desolventizing is accomplished by heating and acting with live steam. Steam consumption is of the order of 1 kg per 12–14 kg of the processed rapeseed. Removal of the solvent remnants from the rapeseed meal is quite difficult, therefore on leaving the desolventizer it still contains 300–1200 mg/kg of solvent in the case of hexane (Beach, 1983). Usually the content of 1000 mg/kg is considered satisfactory (Ward, 1984). A large content of solvent corresponds to a high fat content. Humidity exceeding ca. 25% is reduced to 12–14% either in the further part of the desolventizer, or in a separate drier. The meal passes subsequently through an air cooler (where the water content drops below 12%) and is ground in order to remove the lumps sometimes formed. Export meal is often pelletized.

A traditional desolventizer-toaster has the form of a vertical cylinder, divided with plates into (usually) 6 sections. The meal moved by stirrer blades falls from one plate to another and is steam treated on its way. Solvent and water vapours are passed through an impurity trap to a condenser. The meal passes next through a cooling section and leaves the apparatus through the bottom. Since the meal leaving the desolventizer has often a temperature of 110°C and contains 20% of water, an additional vertical heater is used, similar to the desolventizer-toaster, yet without the steam supply. Instead, it is equipped with a device for venting the meal. Drying and cooling can also be accomplished using a fluidized bed.

Another solution is the Schumacher apparatus, acting as a desolventizer-toaster-drier and cooler. It is equipped with perforated plates supplying steam to the hexane removal section, hot air to the drying section and cool air to the cooling section (Niewiadomski, 1979).

Solvent extraction without prepressing. This method has been used up till now for raw materials low in fats, e.g. soyabean. Recently it is also used for processing raw materials rich in fats, e.g. rapeseed. The primary advantage of this method is a decrease in the capital and operating costs connected with the application of the expellers. On the other hand, the main disadvantage is the difficulty in obtaining meal of a sufficiently low oil content. It usually contains slightly more fat, hence the yield of oil is lower. In certain cases this is compensated by the higher price of meal, containing more fat (Niewiadomski, 1983).

Till now the results obtained during direct solvent extraction of rapeseed are not better than in the case of prepressing coupled with solvent extraction.

1.1.2.3 Solvent losses

Solvent losses during extraction of fat have not only economical meaning, since they increase the fire hazard and the pollution of the atmosphere surrounding the extractor hall. Due to this, numerous investigations are carried out on decreasing these losses. In the seventies the losses of hexane during the industrial solvent extraction of soyabean were equal to ca. 1.25 kg per 1 t of the meal on average. From this amount ca. 0.5 kg/t remained in the meal and liquid wastes, therefore 0.75 kg/t escaped in gaseous form. This amount was distributed in the following way: losses during meal drying 0.55 kg/t, normal leaks 0.05 kg/t and losses during miscella distillation 0.15 kg/t (Krauze, 1978). In the eighties the solvent losses are 0.15% on average with respect to the mass of the raw material, and it is possible to decrease them down to 0.06% (Heilman, 1983).

The greatest losses are due to mechanical reasons, still they can be reduced by avoiding shutdowns and assuring the tightness of the entire equipment. The second position is occupied by losses during meal desolventation. Hexane content of the meal leaving a desolventizer-toaster varies between 0.01% and 0.3% (Myers, 1983). In a plant processing 1000 t/d this amounts to 536 1/d. In third place are the losses in oil. The ignition point (i.p.) of the oil depends on its hexane content. When this content is equal to 0.1%, the i.p. is equal to 120°C, while at 0.05% content i.p.=160°C. In the next place are losses in the air leaving the equipment. The most sensitive method of detection of hexane in this air is smell, which can be detected already when its content exceeds 500 mg/kg. Losses can also occur in the liquid wastes, yet a properly operated evaporator, in which the liquid wastes are boiled for 5–10 min., and their temperature at the outlet is at least 92°C, allows minimization of these losses.

In order to decrease the hexane losses it is necessary to observe certain production conditions. The basic requirement is the continuity of the extraction process. A liquid wastes evaporator should assure the recovery of the smallest traces of the solvent, and the volume of the settling tank should be 1.5 times larger than the maximum amount of miscella contained in the equipment (Sandvig, 1983). The requirements of plant safety are very similar to that of environment protection, since evolution of an inflammable and toxic solvent outside the apparatus, and next to the building, endangers both work safety and the environment. The indicators of hexane presence can be

Fig. 1.7 Treatment of extraction meal (Becker, 1983): 1—conveyor, 2—desolventizer 3—drier, 4—cooler, 5—aspirator.

the control of its content in the surrounding air, the temperature of liquid wastes leaving the evaporator, the smell of meal leaving the desolventizer, etc.

Meal leaving the extractor passes through the following apparatuses: desolventizer-toaster, drier and cooler (Fig. 1.7). Fumes are carried away from this equipment in three places. All the outlets are connected to the solvent recovery installation. In spite of this, uncondensed vapours cause an unpleasant smell of the surrounding air. In some designs all the operations

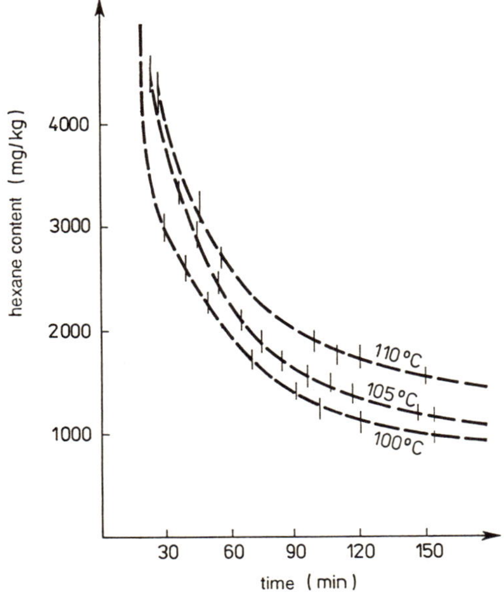

Fig. 1.8 Desorption kinetics of rapeseed meal in a desolventizer-toaster depending on temperature (Wolff, 1983).

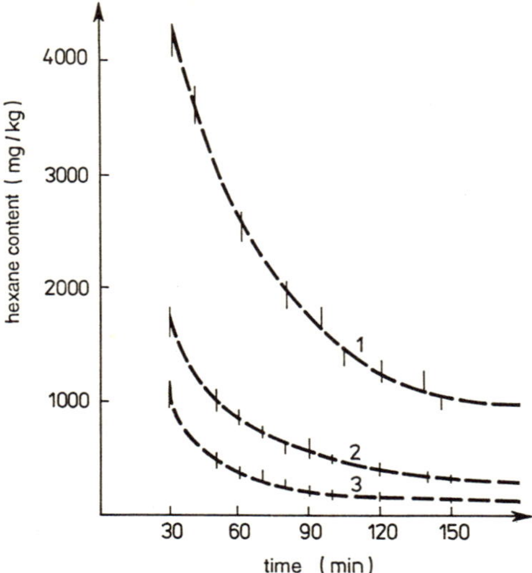

Fig. 1.9 Residual hexane content in meal depending on time of heating the seeds at 70°C (Wolff, 1983): 1—unheated seed, 2—seed heated at 6.2% humidity, 3—seed heated at 4.5% humidity.

take place in one apparatus (Becker, 1983). Such a solution yields lower solvent content in the meal and is simpler in operation. Meal from such an apparatus has a temperature close to ambient.

The content of solvent remnants in the meal is an important consequence of the solvent extraction process and basically depends on the residence time of the meal in the desolventizer-toaster. However, this time should be limited, since its prolongation causes deterioration of the protein — the basic component of the meal. This dependence is presented in Fig. 1.8. It follows from this figure that an increase in the temperature does not basically influence the solvent content of the meal (Wolff, 1983). It has been established that three other parameters play a more important role: heating of the seed before flaking, dehulling and shortening of the contact time of hexane with flakes. The dependence of the amount of hexane in the meal on the time of heating of the seed at 70°C, reducing the soyabean humidity from 7.5% to 6.3% or to 4.5%, is illustrated in Fig 1.9. The decrease in hexane content is much faster than the loss of humidity. It seems that the reduction of the hexane content is due to changes at the seed surface taking place during heating. The results of the investigations on the influence of the hull content in rapeseed are depicted in Fig.1.10. It follows from this figure that the effect of hull on the residual hexane in the meal is distincly adverse.

The last parameter significantly influencing the amount of hexane in the meal is the contact time of hexane with seed (Table 1.1). Although this effect is clear, yet this parameter also determines the content of the residual fat in the meal, which is also significant. An accurate calculation can reveal to what

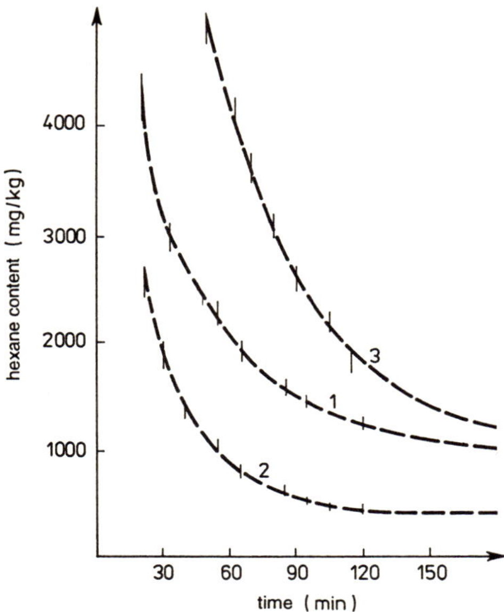

Fig. 1.10 The effect of hull content in rapeseed on the residual hexane content in meal (Wolff, 1983): 1—meal from non-dehulled seed, 2—meal from dehulled seed, 3—meal from hulls.

extent the fat should be extracted from the seed, taking into account not only the refining losses, but also the solvent losses.

Table 1.1 The effect of rapeseed extraction time on the residual hexane content (Wolff, 1983)

Residence time of meal in the desolventizer-toaster at 105°C (min)	Residual hexane content (mg/kg)			
	6 h extraction		2 h extraction	
	not dehulled	dehulled	not dehulled	dehulled
60	2140	1000	1180	740
90	1540	600	740	380
120	1220	460	560	260

The hydrocarbon content of the gaseous phase over the desolventized meal is determined by a relationship arising from the Freundlich law:

$$x = k \cdot c^n \quad \text{or} \quad \log x = \log k + n \cdot \log c$$

where x —amount of hydrocarbons absorbed by the meal, c —concentration of hydrocarbons in the gaseous phase.

Since n is close to 1, the course of absorption can be described by a single parameter, called standard absorption (k). Its value is proportional to the fat content of the meal, therefore the more fat left in the meal, the more solvent in the gaseous phase over the meal, which increases the explosion hazard.

Water vapour from miscella distillation and meal desolventizing contains apart from the solvent, also air introduced by the oil cake. Moreover, a certain amount of air gets in through leaks in the vacuum system. Fumes from the terminal condenser consist of air saturated with water and solvent vapours. Solvent vapours are trapped in a packed tower, where they flow countercurrently to a mineral oil spray. This oil is subsequently heated, deprived of hexane and cooled. The hexane content in gases leaving the ventilation system should be lower than 1.3% vol., i.e. below the explosion limit. Usually solvent losses during rapeseed processing are much greater than in the case of soyabean processing.

Solvent recovery. Basically four methods of solvent recovery are used, viz. indirect condensation, absorption, adsorption and direct condensation.

1. Indirect condensation. Its main disadvantage is the necessity of applying a very low temperature if the solvent content of air is lower than 10 g/m^3. Therefore this method is suitable for fumes containing large amounts of gases for recovery.
2. Absorption. It requires subsequent desorption, which is connected with a substantial development of the instrumentation and the expensive use of a means for solvent stripping.
3. Adsorption. At least two absorbers operating alternately are required. It is vital that the inflammable gases are diluted to a large extent, so that their concentration is lower than 25% of the lower explosion limit.
4. Direct condensation. Compared to the remaining methods it is the most economical and does not pollute the wastes. Its running costs are ca. 1/3 of that of adsorption (Nitsche, 1984).

The thermal energy consumption necessary for the removal of solvent from the meal leaving the extractor, while leaving its humidity at a standard level of 12% and the temperature at 30°C, depends on the humidity of the flakes and the solvent content. In the case of soyabean the humidity of the meal cannot be reduced below 18%, since these are the requirements for suppressing the urease activity and obtaining the anti-trypsin agent. Moreover, decreasing the humidity is limited by the need to use the heat of the vapours leaving the desolventizer for the first stage of miscella distillation. These parameters determine the limits of application of preliminary solvent removal. There are a few methods of such a preliminary desolventizing. They can be divided into mechanical ones, thermal ones and ones incorporated in the system including partial desolventizing. Figure 1.11 presents a possible system of preliminary desolventizing. The choice of the most proper method depends on the extraction system and the steam consumption in this system. For example, preliminary desolventization can be accomplished by a prolonged time of bleeding in the extractor or in an additional zone, by a removal with a heated worm, by pressing the humid meal,

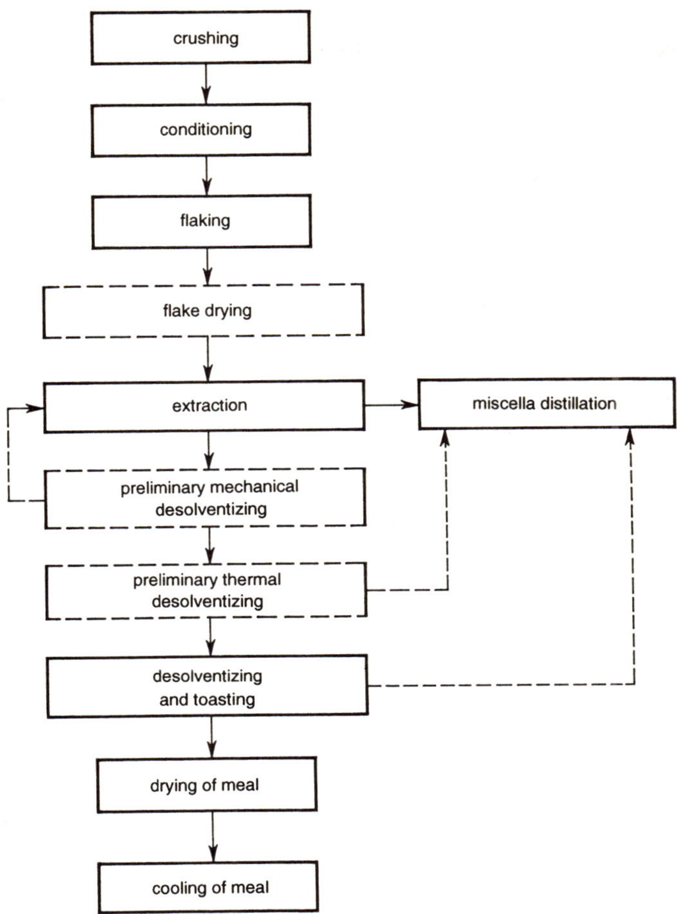

Fig. 1.11 Preliminary desolventizing of seed (Weber, 1983).

etc. The savings in the industry vary depending on the applied method from 15.5 to 145.5 kg of steam per 1 t of the raw material (Weber, 1983).

1.1.2.4 Application of supercritical fluid

The solvents used for solvent extraction of the oil seeds are most often extraction naphta or hexane. However, the prices of these hydrocarbons increase, and the limitations in their occupational safety due to toxicity, inflammability and environmental hazard promote a search for other solvents. From seventy possible solvents the most proper ones seem to be ethanol, methylene chloride, isopropanol, aqueous acetone solutions and mixtures of hexane, acetone and water. Depending on the kind of the solvent, the composition of the meal, the wastes and the fumes change. The possibility of application of

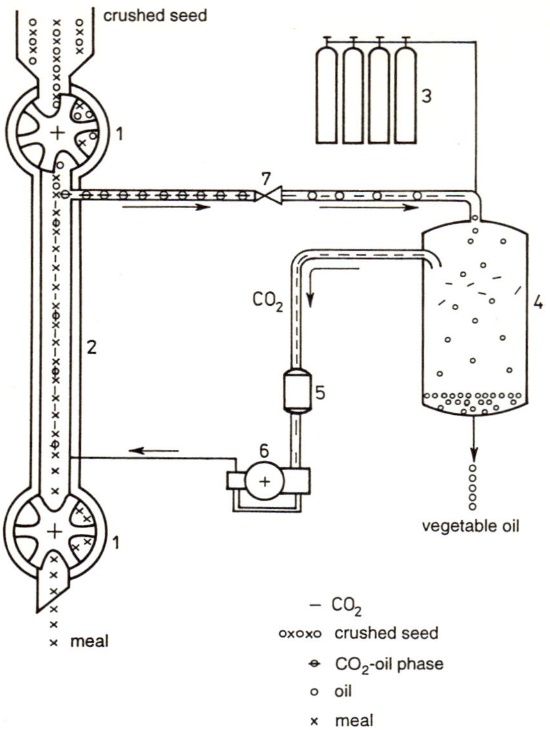

Fig. 1.12 Experimental apparatus for extraction of fat from seed with supercritical carbon dioxide (Mangold, 1983): 1—rotary lock, 2—extractor, 3—CO_2 cylinders, 4—CO_2 separator, 5—filter, 6—compressor, 7—expansion valve.

isopropanol and methylene chloride was recently investigated. These solvents cause simultaneous separation of undesirable components, like aflatoxins, gossypol and alcaloids, and a removal of the specific smell, which enables wider application of meals in animal and possibly human nutrition. Reducing the health and fire hazard, easier dissolution of neutral fats and reduced solution of phospholipids, FFA and non-toxic pigments, can constitute additional advantages of the new solvents (Johnson and Lucas, 1983).

Research is also carried out on application of liquefied gases or supercritical fluids for this purpose. Carbon dioxide is one of the suitable gases (Mangold, 1983). Its boiling point at 100 kPa pressure is equal to –78.5°C, critical pressure 7.29 MPa, while the critical temperature is +31.3°C. Above the critical point the CO_2 is in supercritical state. Apart from CO_2, other compounds can also be used as supercritical solvents, e.g. ammonia, nitrogen oxides or water. However, CO_2 is particularly suited for the extraction of foodstuffs, since it is non-toxic and non-flammable. The technological outline of such an extraction is depicted in Fig 1.12. The gas is compressed to 35 MPa and flows through the ground seed. The oil, extracted as two fractions, is obtained by reducing the pressure. The first fraction is separated at 20MPa, while the second — at 3–6.5MPa, a pressure lower than the critical one. The liberated gas is again compressed and recycled to extraction. Only traces of phospholipids pass to the oil when using CO_2 as the solvent. Therefore the oil is free from lecithin and other polar lipids. The meal preserves the normal composition of the protein substances. Using this method it is possible to obtain with a good yield lipids less polar than triglycerides, e.g. waxes from jojoba seed.

The use of supercritical CO_2 as solvent has been investigated for a number of seeds (Friedrich *et al.*, 1982; Friedrich & Pryde, 1984; List *et al.*, 1984(a) and 1984(b); Fattori *et al.*, 1988). In general, the oils had improved quality characteristics as compared with hexane extraction, reducing the losses and the costs of subsequent refining processes. Cottonseed oil was lighter in colour, easier to bleach and required only 0.1–0.2% excess of 10% lye in refining, leading to lower losses. Extraction was at 800 p.s.i. and 50°, and wet hulled corn germ oil extracted under the same conditions required lower levels of bleaching earth. Soyabean oil was lighter in colour and had a lower iron content. All the oils had very much lower phospholipid contents, and for soyabean oil the degumming step, with its attendant oil losses could be omitted.

The main problem in introducing supercritical CO_2 as a solvent in oil production is the proper design of the seed inlet and meal outlet, taking into account the high pressure in the extraction zone. Application of another solvent would result in a change of the composition of the extraction wastes and the meal.

Apart from application for the extraction of fats from seed, CO_2 can also be used for refining and obtaining the by-products. Supercritical CO_2 reveals very high selectivity of triglycerides dissolution compared to phospholipids dissolution, which enables pure lecithin to be obtained (Mounts, 1983). Application of supercritical CO_2 for deodorization of oils makes it possible to discontinue the use of steam, which always causes the hydrolysis of fats.

A column of 15 m height and 6 cm internal diameter, heated to 90°C, was used in the experimental equipment. The separator was heated to the same temperature. The productivity of the column was 5 kg of oil per hour. A deodorized oil of low FFA content was obtained. Owing to the utilization of different solubilities of the accompanying substances, like tocopherols, sterols, hydrocarbons, pigments and FFA, it becomes possible not to use steam deodorization, thus avoiding atmospheric pollution with the deodorization fumes. At the same time, by applying proper temperature and pressure a method of fat refining without any chemical processes involved can be developed.

1.1.2.5 Other methods of the extraction of fats

Apart from pressing or solvent extraction, it is also possible to extract fats by hydrolysis of seed. Laboratory scale investigations have allowed develop-

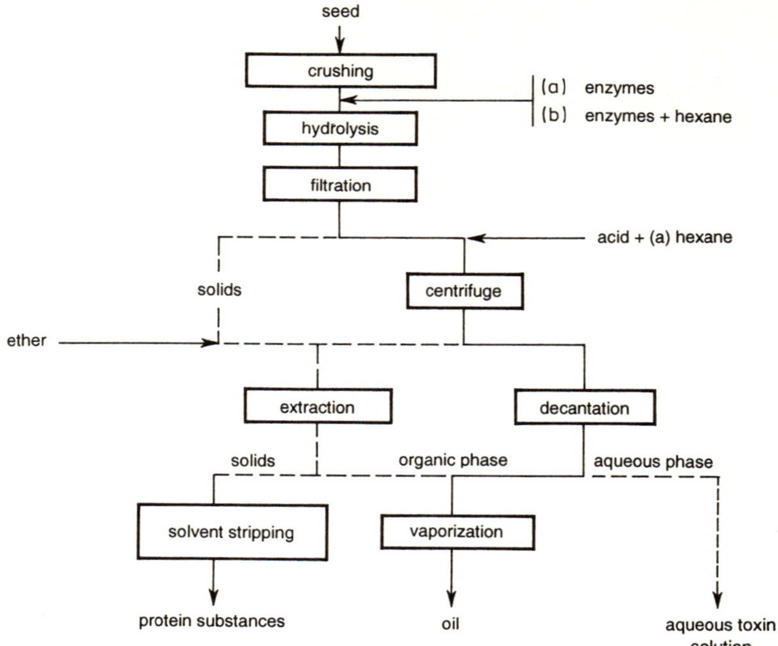

Fig. 1.13 Outline of the hydrolysis-extraction process (Fullbrook, 1983).

ing a technology, the outline of which is presented in Fig. 1.13 (Fulbrook, 1983). In order to utilize the activity of enzymes, the process was carried out for 4 h at a temperature of 50 and 60°C. To inactivate the enzymes, the seeds were heated for a short period to 80°C prior to the separation of fractions or solvent extraction. Introduction of hexane significantly increases the yield. Depending on the applied enzyme and its concentration, using hexane it is possible to extract nearly 90% of fat contained in soyabean and 72% of that contained in rapeseed. This method (verified so far on a laboratory scale) yields oil of such purity that refining is unnecessary. In the case of rapeseed, phytic acid and other toxins are removed from the meal and pass to the wastes in the form of an aqueous solution.

One method of improved extraction of oil from rapeseed, sunflower seed, linseed, etc., is the VPEX system (Homan *et al.*, 1981). It is based on the elimination of crushing, flaking and conditioning of seed before the preliminary pressing. Also in the case of direct solvent extraction, i.e. without preliminary pressing, the VPEX system comprises a smaller number of operations (Fig. 1.14). Part of the oil is pressed by the EP-01 press, which leaves 19–22% of the oil in the rape cake, the temperature of the seed being 10–20°C, and the water content 7–9%. The cake extracted at 60°C and 10–12% humidity liberates residual fat at a rate similar to that achieved in the traditional method, yet higher than in direct extraction.

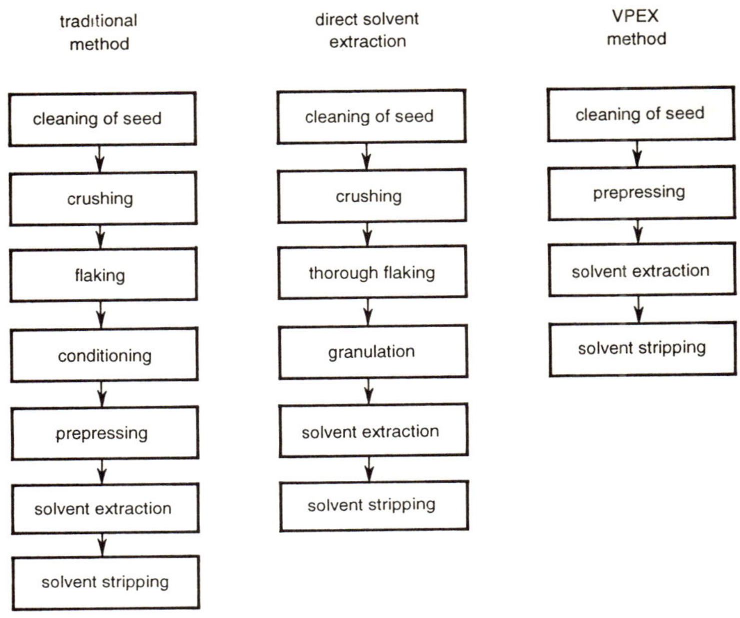

Fig. 1.14 Methods of solvent extraction of edible fats (Homan *et al.*, 1981).

1.1.3 Processing of palm fruit

The oil palm (*Elaeis guineensis*) has within the last 30 years become the world's second most important oilseed crop, with a production of palm oil in 1991 of 11.3 million tonnes and additionally of palm kernel oil of 1.4 million tonnes. While the processing of palm kernels is similar to that of the oilseeds already discussed, palm oil is obtained from the fruit flesh and its extraction and the waste products are quite different. Palm fruit is borne in large bunches weighing 30 kg or more and containing more than 1000 individual fruits. The bunches are harvested by cutting the whole bunch from the tree and are brought into a processing mill within a few hours. Early processing is essential to prevent biodeterioration, resulting in high free fatty acids and loss of bleachability. The processing sequence is shown in Fig. 1.15, which also shows the waste streams produced (Berger, 1983).

1.1.3.1 Sterilization of fruit

On arrival at the oil mill, fresh fruit bunches (FFB) are loaded into cages placed on bogey wheels running on rails. Each cage holds 2.5 tonnes of FFB and up to seven cages are pushed on the rails into a horizontal pressure cooker. A typical cooking operation takes a total of 60–70 minutes at a press-

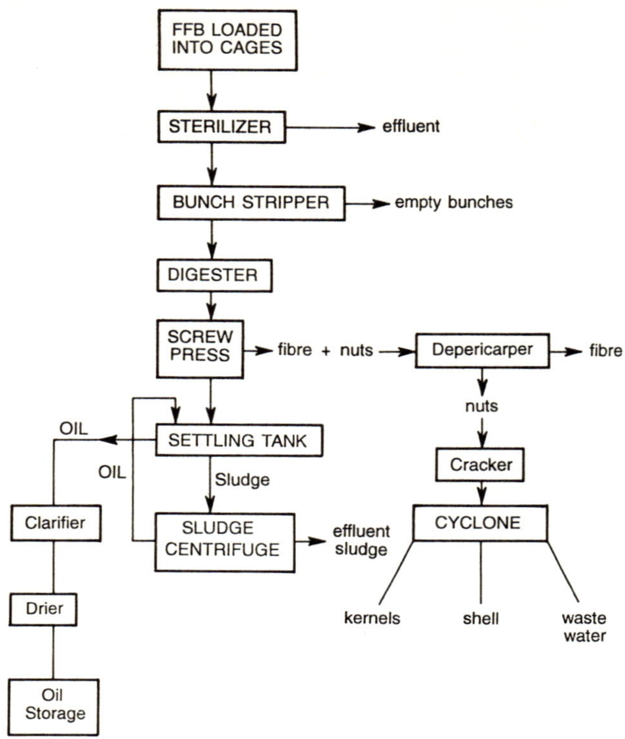

Fig. 1.15 Oil mill process flow sheet.

ure of 3 bar. In order to displace entrapped air it is usual practice to bring a charge up to pressure and release pressure twice during the cooking period. The effluent from the sterilizer contains some oil and both dissolved and suspended matter (see Table 1.2). The purposes of cooking are:

(a) to inactivate lipolytic enzymes,
(b) to enable the fruits to be detached easily from the bunch stucture,
(c) to soften the fruit flesh for subsequent processing,
(d) to loosen the kernel inside its shell.

1.1.3.2 Stripping of fruit

After removal from the cooker, each fruit cage is lifted by crane and the contents emptied into a hopper which feeds the fruit stripper. This consists of a cylindrical cage about 2 m in diameter whose walls are metal bars with gaps between them. The cage rotates about the horizontal plane and baffles lift the fruit bunches and allow them to drop to the bottom. The fruit is knocked out and drops through the bars onto a conveyor. The empty fruit bunches form a second waste stream.

1.1.3.3 Digester and screw press

The fruit is fed into a digester, consisting of a vertical cylinder fitted with rotating beater arms and steam heated. The fruit is thoroughly disintegrated and the oil cells in the flesh are ruptured. The mashed fruit drops into a continuous screw press. The liquid stream from one end of the press consists of oil and water with finely divided and dissolved fruit solids. At the other end of the press relatively dry fruit fibre is extruded together with the nuts. The fibre has a residual oil content of 7% (dry weight basis).

1.1.3.4 Settling and clarification of oil

The liquid stream from the press passes through a vibrating screen into a settling tank with 2 h residence time, where it is diluted with water to aid separation. Oil from the upper layer is passed through a clarifying centrifuge, dried and pumped to storage. The lower sludge layer is passed through a self-cleaning sludge centrifuge. The oil phase is recycled to the settling tank, while the lower layer is a waste stream.

An alternative method for treating the oil stream from the screw press depends on the use of either two-phase or three-phase decanter centrifuges (Lim, 1988). It has the advantage of reducing the amount of water used and both the volume of waste water and its BOD by at least half. The solids waste discharged from the decanter may be dried using waste heat from the boiler flue gases and used as a fertilizer. Its content of mineral components is N 1.87%, P 0.46%, K 2.35%, Mg 0.99% and Ca 0.067%.

1.1.3.5 Recovery of palm kernels

The cake of solids extruded from the screw press contains fibres and nuts. It is first broken up in a trough conveyor fitted with helically disposed rotating beater arms. Most of the fibre is separated pneumatically. The nuts are then tumbled in a rotating drum to remove residual fibre, conditioned in a hot air stream to 11% moisture, and cracked in a centrifugal cracker. A rotor hurls the nuts against a hard metal cracking ring. Shell fragments are separated from kernels by gravity, usually using a pneumatic column followed by a hydrocyclone. The kernels are dried to below 8% moisture in a hot air stream and eventually processed into oil and meal by conventional pre-pressing and solvent extraction processes. The fibre and some of the shell are used to fire the oil mill boiler, and provide all the steam and electrical energy required by the mill.

The empty fruit bunch stalks may also be used to augment the boiler fuel, in which case they are first cut up, and partly dried by pressing. Alternatively they are burnt in an incinerator and the ash used as a fertilizer. The analysis of the ash is: K_2O: 41.4%, P_2O_5: 3.7%, MgO: 5.8% and CaO 4.9%. A third option, avoiding the atmospheric pollution of the incinerator, is to return the empty bunches to the field as a mulch, when

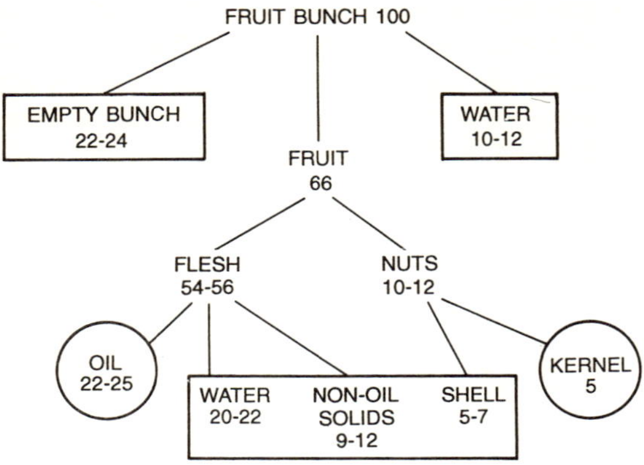

Note. Circles — products.
Square brackets — waste.

Fig. 1.16 Yield of oil mill products.

the fertilizer content is released slowly over three years. The value of empty bunches as fertilizer is indicated by the following figures (Hornus, 1992): 1 ton of empty bunches equals 3.4 kg urea, 2.3 kg natural phosphorus, 11.7 kg KCl and 2.7 kg kieserite.

1.1.3.6 Yield of oil mill products

Quantitative data for the various products of the milling operation are given in Fig. 1.16.

The composition of the various liquid effluent streams indicated in Fig. 1.15 is shown in Table 1.2.

Table 1.2 Composition of oil mill effluent

	Sterilizer		Oil clarifier		Hydrocyclone		Mixed	
Volume, in tons per ton oil produced	0.9		1.5		0.1		2.5	
Oil, %	0.4	4.3	0.75	5.25	0.03	0.11	0.6	4.5
Dissolved solids, %	3.4		2.2		0.01		2.1	
Suspended solids, %	0.5		2.3		0.07		1.8	
pH	5.0		4.5		-		4.7	

1.1.4 Processing of olives

The cultivation of olives is an age-old tradition in the countries of the Mediterranean basin, and today more than 90% of the world production of

1.5 million tons annually still originates in this region. Olive oil is a fruit flesh oil, and the process of extraction has some similarities to that for palm oil, but it is simpler. After washing, the fruit is crushed, traditionally in a type of end runner mill. The solid matter is separated in a sludge centrifuge. The oil in the liquid stream is clarified in a second centrifuge. The solid matter is dried and solvent extracted to yield a lower grade oil. A flow diagram, adapted from Annesini (1983), is shown in Fig. 1.17, which also gives the material balance.

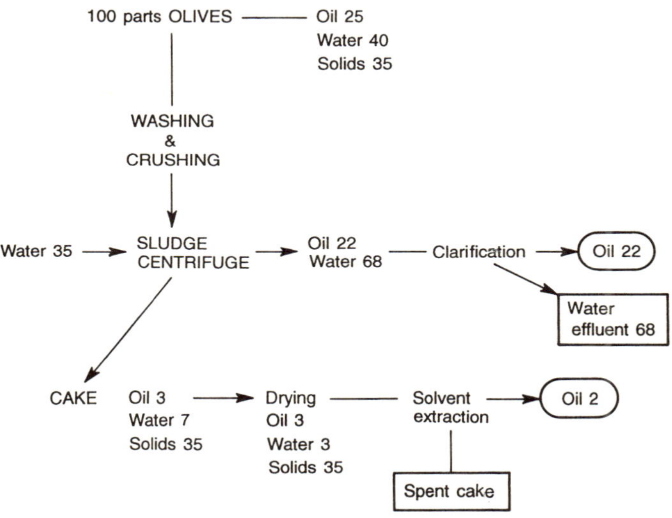

Fig. 1.17 Olive oil extraction.

The effluent water contains 15% organic compounds and 1.0% inorganic salts. It has a BOD 40,000–90,000 ppm, a COD 50,000–100,000 ppm, suspended solids 40,000–80,000 ppm and pH 4–5 (Annesini, 1983).

1.1.5 Degumming of oils

This process aims not only at the removal of gums, consisting mainly of phospholipids, but also at the utilization of the lecithin contained in them. Another goal of degumming (hydration) is preventing the formation of cumbersome sediment during prolonged storage or transport. This is the reason for rejecting non-hydrated soyabean oil for transport by tank ships.

The phospholipid content of oils varies significantly according to the kind of raw material. Fig 1.18 shows that it is highest in soyabean. Due to this, soyabean lecithin covers almost the total demand. Rape is medium rich in phospholipids. The content of phospholipids in crude palm oil can vary from 10 to 35 mg/kg in terms of phosphorus (Maclellan, 1983). Complete degumming involves the removal of both hydratable (HPL) and non-hydratable (NPL) phospholipids. The former process is usually carried out in the extraction plant, while the latter — usually called desludging — in the refinery. However, due to their interconnection they will be discussed in the same paragraph.

1.1.5.1 Degumming of soyabean oil

Degumming, as well as alkali refining and wax removal, is used selectively for various kinds of oil depending on their chemical composition (Table 1.3). Following extraction, the oil of a temperature similar to that used in hydration (50–70°C) is subjected to the action of hot water or steam.

Table 1.3 Refining processes used for some oils (Haraldsson, 1983)

Oils (fats)	Degumming	Alkali neutralization	Dewaxing
Various*	(+)	+	–
Sunflower, maize	(+)	+	+
Soyabean	+	+	–
Palm and animal tissue fats	–	(+)**	–
Fish	–	+	–
Rice	+	+	+

* E.g. peanut, rapeseed, sesame.
** Those of good quality can be physically neutralized.

An addition of 0.5% of water is sufficient to precipitate the HPL from soyabean oil, yet 2% is used in practice in order to facilitate the separation

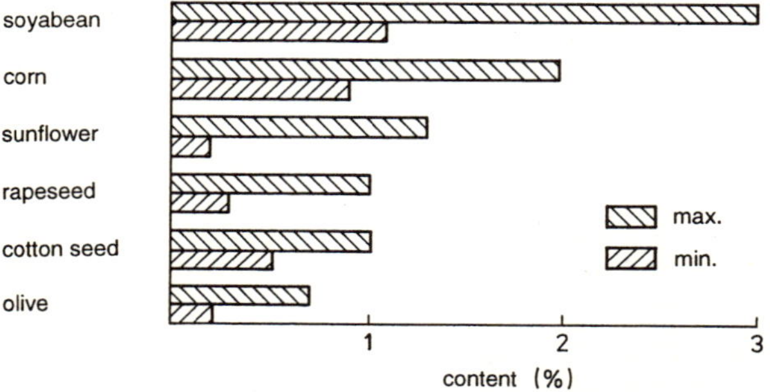

Fig. 1.18 Phospholipid content of some vegetable oils (Haraldsson, 1983).

of the lecithin phase from the oil phase. On the other hand, too large an excess of water results in accelerated fermentation of gums. Occasionally part of the steam used for the removal of the hexane residues from the oil is introduced into the hydrated oil. However, the determination of the amount of steam necessary for the precipitation of phospholipids is not easy and therefore the addition of hot water is used more often.

Hydrogen peroxide or benzoyl peroxide are used already at the stage of hydration in order to brighten the colour (Haraldsson, 1983).

The degumming temperature should be equal to 50–60°C, and the contact time with vigorous mixing can be limited to a few seconds, provided that the process is followed by a few minutes' settlement (Pardun, 1983).

Application of higher temperature lowers the hydration degree, since the solubility of phospholipids in oil increases. On the other hand, a too low temperature increases the viscosity of the oil, which hinders separation. Due to the presence of surfactants, emulsions can easily be formed. To avoid this phenomenon it is necessary to be cautious in starting the supply pump and the inlet to the centrifuge. Wet gums can either be mixed with the meal, or dried in a thin film drier before further refining.

Typical crude soyabean oil provides 3.5% of gums, consisting of 25% of water (Swern, 1964) and 75% of the oil phase. This phase contains 1/3 of oil and 2/3 of substances insoluble in acetone.

A process properly carried out yields oil with a low content of residual phospholipids and lecithin with a low content of oil. It is possible to obtain oil containing less than 0.1% of phospholipids, as in the Alcon process (see Fig. 1.6). The amount of water necessary for hydration in this case is the same as in the traditional process or even smaller, therefore no drier overloading occurs. An important feature of the Alcon process is its precluding the action of the enzymes, i.e. lipoxygenase and phospholipase, which cause the formation of NPL. The sediment is centrifuged in a disk-type centrifuge.

The result of hydration depends primarily on the following conditions:

- Storage of flaked seed prior to extraction for more than 24 h significantly deteriorates their quality. In particular, the content of NPL, FFA, AV and PV increases, which is due to the action of enzymes (lipoxygenase) on phospholipids.
- An increase in the water content in the extracted soyabean increases the enzymatic activity. Although it has not been precisely established how much NPL is formed during extraction, the results obtained so far indicate that at least 0.05–0.22% is formed (Kock, 1983).
- Increasing the extraction temperature within the limits of 40–60°C results in a considerable increase in the PL content in the oil after degumming.

Application of a modified soyabean pretreatment causes changes in PL composition, viz. the content of phosphatidyl choline increases while that of phosphatidyl ethanolamine decreases (Table 1.4). These initial operations

Table 1.4 Soyabean lecithin phospholipid content (Kock, 1983)

Phospholipids	Content (%) in lecithin from flakes	
	prepared in a traditional way	preliminary moistened and heated
Phosphatidyl choline	33	46
Phosphatidyl ethanolamine	30	23
Lysophosphatidyl choline	2	2
Monophosphatidyl inositol	14	8
Phosphatidic acid	17	19
Other phospholipids	4	2

result in an increase of the amount of the lecithin obtained by ca. 100%. A good, properly degummed oil often contains only 4–5 mg/kg of phosphorus; in such a case 0.2% of bleaching earth is sufficient. Furthermore, loss of bitter taste of the meal due to such a treatment of the flakes is also an important advantage. Experiments have proven that seed heated up to 90°C and losing 2% of water yields after extraction an oil containing less NPL compared with oil from unheated seed. Similarly the AV was lower by 0.12 to 0.32. The selection of optimal parameters for the entire process of soyabean oil extraction makes it possible to obtain an oil of a very low NPL content after hydration, equal on an industrial scale to 0.03% PL, which corresponds to 10 mg/kg phosphorus content. A reduction of the NPL content is accompanied by an increase in the amount of the lecithin obtained. It is in this way that an extraction plant increased the annual yield of lecithin from 0.6% to 1.2%.

Another method of thorough degumming is the application of two separators in series. The temperature of oil is 90–95°C , while that of water is 20–25°C. The back pressure in both the separators varies from 2.4 to 3.0 MPa. For example, when the phosphorus content before the first separator is 3.5–4.5% and the water content is 2–3%, they are reduced at the outlet from the second separator down to 0.07–0.09% phosphorus and 0.2–0.6% water (Lurgi system). Although nearly all the investigations concern soyabean oil, yet it can be assumed that in the case of rapeseed oil the results would be similar. It can generally be stated that under the conditions of extraction the lipoxygenase causes the oxygenation of the phospholipids, due to which they form complexes with soyabean meal proteins. As these complexes are insoluble in hexane, they decrease the lecithin yield (Kock, 1983). A reduction in effluents and an improvement in refining yield are obtained in the 'total degumming' process described by Dijkstra and Van Opstal (1989). In the first step, dilute phosphoric or citric acid is mixed very intimately into the warm oil and allowed to react about two minutes to effect decomposition of metal phosphatide complexes. Reaction products are removed by partial neutralization of the added acid, avoiding soap formation. Most of the gums, containing little oil, are separated in a first centrifuging. Residual gums are removed in a second centrifuge. These oil-rich gums are recycled. The oil is freed of inorganics by two water washes and dried. It can be chemically or physically refined. The soap stock from chemical refining is free of phophatides and contains only 10–15% neutral oil.

Soya, rapeseed and sunflower oil can be processed to less than 2 ppm iron and less than 5 ppm phosphorus before bleaching.

If the oil is to be stored after hydration, it is dried by spraying in a vacuum chamber, where at the pressure of 7 kPa and at 80°C the water is evaporated down to 0.1%. The oil is subsequently cooled down to 39–49°C and directed to storage.

Degumming by means of acetic acid anhydride is based on introduction of 0.1% of the anhydride to the oil heated to 60° C for 15–30 min and subsequent stirring for 30 min with 1.5% water added. In the case of good quality crude oils the neutralization can be unnecessary, hence the losses connected with this process can be avoided. The lecithin yield is higher, and the equipment for soapstock splitting is needless. This method is nowadays used only for increasing the degumming degree before alkali neutralization (Brekke, 1980a).

1.1.5.2 Degumming of rapeseed oil

Hot oil is cooled down to 82°C and hydrated by an addition of 1–2% of hot (80° C) water condensate (best of all in the amount equal to the content of phospholipids in the oil). Sometimes a small amount of phosphoric acid is added in order to facilitate the precipitation and coagulation of phospholipids. Such a procedure is commonly employed when degumming is accompanied by neutralization. The resultant degumming degree reaches 150–250 mg P/kg, depending on the properties of the rapeseed, storage time, conditon of seeds and the manner of carrying out the process. The phospholipids are precipitated, centrifuged and pumped to a desolventizer, where they are mixed with the meal. They reduce the dusting of the meal and act as a binder in case of pelletizing (Beach, 1983). Even better results can be obtained by thermal decomposition of phospholipase. For this purpose the seed is heated to 100°C (instead of 85°C) in a vertical toaster. Although the oil from the seed heated in this manner contains more phospholipids, yet water degumming removes them to a level attainable only by acid degumming. This method, however, has certain disadvantages, viz. the concentraction of sulphur compounds in the oil, and sometimes also the fat content of the meal increases.

Degumming of pressed oil. In some oil mills the pressed oil lecithin is not removed, or is removed together with the solvent extracted oil. Investigations aiming at the determination of the amount and quality of the lecithin obtained from both these sources have been carried out. It follows from these investigations that the yield of the lecithin from the pressed oil is 0.33%, and that from the solvent extracted oil 1.34% (Kubicki and Wojnarowicz, 1973).

1.1.5.3 Desludging

Hydration removes from soyabean oil ca. 90% of phospholipids, while hexane extracts about half of the phospholipids contained in the seed. For some refining methods (e.g. distillation or the Zenith method) it is necessary to remove phosphorus more thoroughly, hence desludging by various methods is accomplished. The desludging methods can be divided into methods based on utilization of mineral acids, adsorbents, and other methods. The basic criterion of selecting the desludging method is the required quality of the oil directed to the neutralization. The basic indicator is the phosphorus content (Szemraj, 1973).

Acid hydration. It decreases the phosphorus content by a further 10–50 mg/kg. There are certain modifications of the process, yet some features are common. The oil usually should have a temperature of 60°C. Water and acid are introduced to the oil. The mixture is cooled down to 35°C and left for some time for the reaction to take place. The mixture is separated in a centrifuge, the temperature being raised to 60°C if necessary. Some prescriptions recommend the addition of a coagulant in order to facilitate phase separation, which can be hindered due to the properties of the crude oil, the agent used for the degumming, the manner of stirring and decantation. After acid desludging the phosphorus content is 10–20 mg/kg. The main advantage resulting from such a decrease in the phosphorus content is the decrease of the losses during alkali neutralization or preparation of the oil for physical neutralization. Phosphoric, acetic, citric or maleic acids are used for acid hydration.

Phosphoric acid of 85% concentration is used in an amount of ca. 0.2% at a temperature of 70–90°C. The reaction takes place within one minute at vigorus stirring. Directly after the reaction the oil is neutralized, the amount of the sodium hydroxide added taking into account not only the FFA, but also the phosphoric acid. However, lecithin obtained in this manner is darker than that obtained with water only, which decreases its price.

Desludging of the double zero* rapeseed according to Canadian sources is accomplished in the following way: 2% water or steam is introduced to the oil at a temperature of 70–90°C and stirring thoroughly. The mixture is left for 30 min for maturing. Further improvement in desludging of such an oil consists in application of citric acid or maleic anhydride at a temperature of 40° C, for a contact time of 10 min. Then 2% of water is added and the

* Double zero oil contains less than 5% erucic acid, while the meal contains less than 3 mg of 3-butenylisothiocyanate equivalent per 1 g of fat free dry weight.

mixture is left again for 20 min for the reaction to take place. An addition of 2500 mg/kg of oil results in a reduction of the phosphorus content from 541 to 33.3 mg/kg in the case of maleic anhydride or to 50.0 mg/kg in the case of citric acid. Such a reduction of the content of this element yields the following profits (Diosady *et al.*, 1984).

1. As a large fraction of the phospholipids is recovered in the oil mill, it is possible to add them to the meal, thus increasing its feeding value.
2. Desludging in oil refineries can be omitted.
3. Refining losses due to emulsification are reduced.
4. Soapstock splitting is facilitated owing to a smaller content of emulsifiers.
5. Refinery wastes are cleaner owing to a reduction of fat and sludge content.
6. The oil is suitable for physical refining, since it contains less non-volatile impurities.

Maleic anhydride is toxic, hence citric acid is more suitable.

Application of bleaching earth. Another method of total PL removal is a combination of the traditional degumming with phosphoric acid and application of bleaching earth (Grothues, 1982).

The process is started with degumming with water. The PL left in the oil are removed with 85% phosphoric acid and activated bleaching earth. First 0.1–0.15% of phosphoric acid is added to the hydrated, wet oil and stirred for 15 min. Under such conditions a dense, acid gum is formed from the residual phospholipids. Next 1 to 2.5% of bleaching earth is added under a reduced pressure at ambient temperature. The degassed mixture is stirred and rapidly heated to 170°C. After reaching the final temperature the mixture is cooled down to a temperature required for filtration. The bleaching earth, acting as an adsorbent, removes not only the phospholipids and other side-products, but also almost totally the phosphoric acid.

The upper allowable limit of the phospholipid content before deodorization is 0.015% (Kock, 1983). Therefore if the oil is to be physically refined, the removal of the residual phospholipids with bleaching earth allows a reduction of their content down to the level of 0.01%. Depending on the amount of the adsorbent (0.5–1.5%) and the initial PL content, the final content varies from 0.1 to 0.004%. Such oils are suitable for deodorization or physical neutralization. The US trade standard for the degummed soyabean oil permits 0.02% phosphorus. A quality standard for the degummed double zero oil (Forster and Harper, 1983) sets the following limits:

	%		mg/kg
FFA	< 1	phosphorus	< 30
water and insolubles	< 0.3	sulphur	< 5
neutral acid	> 99	pheophytins	< 10

Superdegumming. In search of a method for the most accurate removal of side-products from soyabean oil, a so-called superdegumming has been proposed (Ringers and Segers, US pat.). It can be applied to oils containing less than 1000 mg/kg phosphorus. Superdegumming incorporates 7 unit processes (Segers, 1982, Segers & Van der Sande, 1990). The method allows a reduction of the phosphorus content to the level below 30 mg/kg, and even a level of 10 mg/kg can be achieved depending on the process parameters. The method has been verified on an industrial scale on soyabean, sunflower, rapeseed, peanut and linseed oils. Superhydration removes almost totally the phospholipids from the soyabean and rapeseed oils. The oil is mixed intensively at 70°C with a 50% citric acid solution added in an amount of 0.3–1.5 g/kg and maintained at this temperature for 5–15 min. Then it is cooled down to 25°C, 15 g/l of water is added and everything is left for 1h. The phases are subsequently separated in a centrifuge. An oil containing 7 mg/kg of phosphorus has been obtained during laboratory experiments on rapeseed oil initially containing 225 mg/kg P. The oil was easily separated from the aqueous phase and did not form emulsions after alkali neutralization.

It has been established that citric acid is a good agent for the precipitation of a large fraction of phospholipids from the 00 oil. Two methods are used for this purpose. In the first method a 50% acid solution is introduced to the oil at 35°C and slowly stirred for at least one hour. In the second method the acid is added to the oil at 60°C and vigorously stirred for a shorter period. Attention should be paid, however, that the stirring does not cause breaking up of the precipitate being formed, since this would hinder its separation in a centrifuge. The amount of the acid introduced is equal to 1000–3500 mg/kg, while that of water is slightly above 0.75%. Phase separation is accomplished by centrifuging. Application of superdegumming avoids the use of phosphoric acid, which shortens the refining time and eliminates the use of sodium hydroxide and sulphuric acid, due to

Table 1.5 Characteristics of liquid wastes from soyabean oil alkali neutralization depending on the degumming method (Segers, 1982)

Components	Superdegumming (No. of experiment)		Traditional degumming (No. of experiment)	
	(1)	(2)	(1)	(2)
FFA (%)	0.44	0.55	0.55	0.67
Phosphorus	18	40	159	231
	Content in wastes per 1 t of oil			
Water (m^3)	0.19	0.19	0.27	0.25
Sulphates (kg)	1.77	2.29	4.00	3.69
Glycerol (kg)	0.11	0.09	0.24	0.33
Fat (kg)	0.11	0.10	0.23	0.29
COD (kg)	0.41	0.43	1.16	1.70
Nitrogen (kg)	2.98	3.04	15.34	24.02

which the wastes contain less salts. Still less contaminated are the wastes from physical refining, since it is sufficient to bleach the superdegummed oils with only 0.7% of bleaching earth at 90–100°C, owing to which the formation of larger amounts of isomers is avoided (Segers, 1982). Table 1.5 presents a comparison of the composition of the wastes from superdegumming and traditional degumming with water. The results of industrial-scale experiments indicate that superdegumming positively influences the quality of the wastes also in the case of alkali neutralization.

A process of refining phospholipid containing oils has been recently patented, in which the oil is degummed with water or water with acid (e.g. phosphoric or acetic) and passed over a granulated active carbon bed. The water is subsequently removed by steam distillation. Oil obtained in such a manner contains less than 5 mg/kg of phosphorus, the PV is lower than 2, and the taste and smell are faultless. The contact time of the oil with the bed is equal to 6–12 h, and the ratio of oil to carbon is 19:1. The bed can be regenerated and used again.

1.1.6 Obtaining of lecithin

Lecithin is a valuable by-product obtained basically from soyabean. It can also be extracted from rapeseed. The quality of rapeseed lecithin is poorer, since usually certain conditions for maintaining quality are not fulfilled (Nieuwenhuyzen, 1976).

Soyabean lecithin is obtained in a manner presented in Fig. 1.19. Gums leaving the centrifuge contain 40–50% of water, which is evaporated in a drier to the level of 1%. Drying can be carried out in a batch vacuum apparatus heated with water of a temperature of 60–70°C. Such conditions

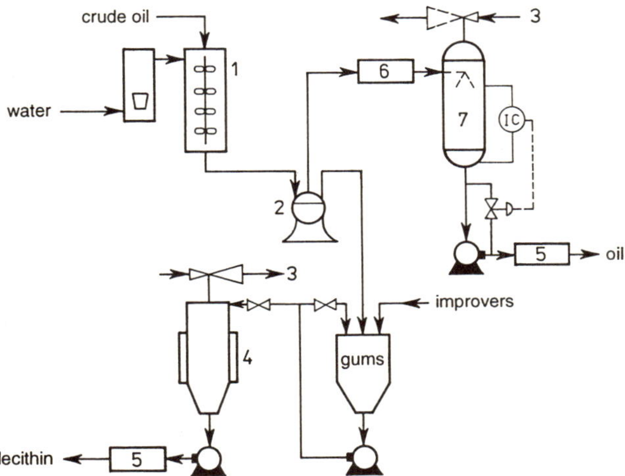

Fig. 1.19 Production of soyabean lecithin (Szuhaj, 1983): 1—mixer, 2—centrifuge, 3—ejector, 4—film drier, 5—cooler, 6—heater, 7—oil drier.

Table 1.6 Parameters of soyabean lecithin drying (Nieuwenhuyzen, 1976)

Parameters	Drier	
	periodic, water heated	continuous, thin film horizontal
Temperature (°C)	60–80	80–95
Time of operation (min)	180–240	1–2
Pressure (kPa)	3–8	7–40

prevent the darkening of lecithin. Continuous devices are usually designed as horizontal or vertical thin-layer evaporators. In the horizontal design the layer is not broken, in contrast to the vertical design. Table 1.6 presents a comparison of the parameters of soyabean lecithin drying in batch and continuous driers. Batch driers are used most often for drying lecithin, although thin-layer driers allow a better control of the process parameters. Drying proceeds well also in a rotary thin-layer drier, in which the lecithin gum remains for 1–2 min at 100°C under a pressure of 6.65–40 kPa. Lecithin easily changes under the influence of heat, hence in order to obtain a good quality product it is necessary to strictly follow the optimal drying conditions. After drying the lecithin is cooled down in order to avoid its darkening.

Finished lecithin is stored in bulk or in drums. On heating the temperature should not exceed 48°C. The storage life at 21°C is usually 1 year. Decreasing the temperature to 0°C reduces the storage life. The drums can be heated with steam in order to facilitate pumping. The conditions described concern soyabean lecithin; most of them, however, can be related also to rapeseed lecithin.

Lecithin is obtained using the following unit operations; mass transfer, centrifugation, filtration and heat transfer (Brian, 1976).

Mass transfer occurs between all the unit processes. Viscosity plays an important role in this process. In the case of miscella it is lower than 1 cP. The most viscous is lecithin containing 2–3% water. The viscosity of wet lecithin after degumming increases during heating to over 1 million cP, and when the humidity drops below 1%, the viscosity decreases down to ca. 30000 cP. For this reason gear pumps are used for the transfer of lecithin.

Disk-type centrifuges are used for the separation of lecithin gums. Crude oil destined for removal of lecithin is usually prefiltered. Dry lecithin can also be filtered. Tiny impurities cause turbidity of lecithin. Filtration of dry lecithin yields a clear product of the best quality. The filtration is accomplished under pressure using horizontal fabric filters, with an addition of an auxiliary substance.

Heat transfer takes place during the entire process. In particular the oil requires heating or cooling before filtration or hydration. The lecithin should be heated prior to centrifuging and cooled subsequently before storage.

The following factors influence the quality and the yield of soyabean lecithin:

—Growth conditions of the plant. Depending on the location of the plantation the composition of lecithin fatty acids changes. As in the case of the glycerides, the further north the plantation is situated, the more linolenic acid the lecithin contains. The weather also plays an important role — the colder the summer, the lower the amount of unsaturated fatty acids. Chemical composition of the soil influences the content of mineral substances in lecithin.

—Crop season. Processing of green, unripe soyabean results in changing the colour of lecithin, reducing its amount and changing the composition of the phospholipids. Less phospholipids and more FFA are obtained from frozen seed. This results in a reduction of the content of the components insoluble in acetone and in a reduction of the lecithin viscosity.

—Seed storage. Soyabean is stored 1–2 years before the extraction. Prolonged storage causes an increase in phosphatidic acids and FFA content, which results in a reduction of the viscosity and of the amount of acetone insolubles. Seed damaged by frost contains less phospholipids and more chlorophyll. Seed which is properly vented and stored under conditions of low humidity constitutes the best raw material for the production of lecithin. A sour or burnt smell of the seed is transferred to lecithin. When the hulls are broken, then light and heat cause an increased oxidation and a decomposition of lecithin.

—Solvent extraction. If the temperature of the desolventized oil is too high or should it be not properly cooled after this process, then the lecithin quality is deteriorated. Temperature should be moderate and should last for as short a time as possible. Pressed oil yields worse lecithin. The high temperature of this operation causes a dark colour which cannot be removed.

—Crude oil storage. The quality of lecithin can be deteriorated as a result of too high oil storage temperature. Best of all it should be stored at ambient temperature. The longer the storage time, the lower should be the temperature. Stirring without aeration prevents sedimentation of the phospholipids at the bottom.

—Degumming conditions. Except acetic anhydride, all the additives facilitating degumming spoil the lecithin. Phosphoric acid causes mineralization of lecithin during drying. Oxalic acid changes lecithin into a toxic substance. Inorganic salts cause changes in physical and functional properties. On the other hand, acetic anhydride reacts with lecithin forming partly acetylated phosphatidylethanolamine.

—Storage of wet gums. In order to protect the gums against deterioration during storage, even for a few hours, diluted hydrogen peroxide solution is added.

—Bleaching. Observation of the rules of the procedure reduces the need for bleaching lecithin. A 3% hydrogen peroxide solution in an amount of 1.5% is used for this purpose. Benzoyl peroxide in an amount of 1.5% is used in order to make the bleaching more intensive. Both these peroxides reduce pigments. A certain amount of peroxides should be left in the bleached lecithin, otherwise it rapidly darkens at 70°C.

Lecithins from oils other than soyabean must be refined in order to improve their taste, smell and colour. Purification or fractionation are applied for this purpose. Purification consists in a removal of the parent oil by extraction with acetone, which does not dissolve phospholipids. Mixtures of acetone with other solvents can also be applied, e.g. with hexane and 2–5% of water. Carbohydrates and substances causing a bitter taste are also removed in such a manner. The purified lecithin contains ca. 2% of water.

Fractionation of lecithin into choline and kephalins is carried out by means of ethanol. A few per cent of a monoglyceride is added to anhydrous lecithin to cause its liquefaction. In a first extractor it is mixed with 20–25% w/w of ethanol and transferred to a second extractor, where at 10–30°C the extraction takes place. After the reaction the mixture is separated into a lighter fraction, containing ethanol and the substances dissolved in it, and a heavier fraction. Both fractions are refined and freed from the solvent, which is recycled to the processing (Liebling and Lau, 1976).

In order to broaden the scope of applications of commercial lecithin, it can be modified by the following processes: enzymatic hydrolysis, hydrogenation, hydroxylation, halogenation, etc.

1.1.7 Refining of oils

Crude vegetable oil usually requires refining. A number of processes and operations, in which by-products and wastes are formed, are carried out for this purpose (Fig. 1.20). Preliminary refining is carried out in the oil mill, and the proper process in a refinery. Impurities contained in miscella and in pressed oil are removed in the oil mill.

1.1.7.1 Removal of suspensions

Methods of removal of the impurities contained in fats in a suspended form can be divided into sedimentation, filtration and centrifuging.

Sedimentation. The method is not very effective and requires a long time. Crude oils normally contain 0.5% water, which usually settles together with the impurities. This creates favourable conditions for the decomposition of fat by enzymes and bacteria. Fat occluded by the impurities rapidly undergoes hydrolysis. As a result FFA are formed, which together with the smell caused by bacteria deteriorate the quality of fats. Thus it is a simple, but expensive and ineffective method. Fully clarified oil cannot be obtained, since stirring occurs during phase separation, which causes renewed partial mixing and creates the necessity of filtration or centrifuging.

Fat can be removed from the separated sediment using centrifuges, and the impurities constitute a waste.

Filtration is a more modern method of clarification. A proper selection of the filtering material, which should stop the impurities and let the oil through, constitutes a significant condition for the effectiveness of the process. The following materials are suitable: cotton and woollen fabrics, artificial fabrics, gauzes and special filtering materials like asbestos, silica, kieselguhr, etc. Filtration is used at various stages of oil refining, yet the principle of the process remains unchanged. Filtration is accomplished by means of chamber and plate-and-frame filter presses. Filter presses of large filtering area and small space for the trapped impurities are used for the filtration of crude oil, containing a small amount of solid components, yet often of colloidal or slimy character. Sediment of such an origin tends to clog the pores of the filter material. Chamber filter presses are therefore used for this purpose. On the other hand, in cases when the amount of the sediment is large, e.g. during separation of bleaching earth, filter presses of large spaces for the trapped material are used — for instance plate-and-frame presses. Clarifying agents are usually used when the sediment occurs in a small amount or has a slimy character. These agents are asbestos fibers, bleaching earth, bentonite, silica and other agents of high adsorptive

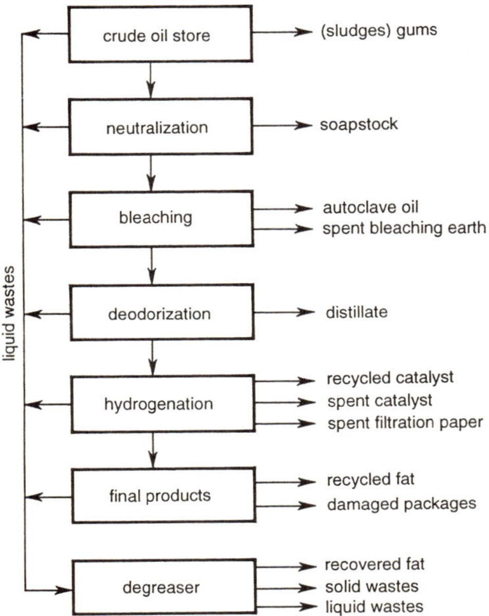

Fig. 1.20 Formation of by-products and wastes in traditional vegetable oil refining.

capacity. A 1% addition is used. Pressure filters are also used. Continuous filters constitute a modern solution eliminating the burdensome attendance of the batch filters. The majority of these filters can be controlled automatically. The path of the sediment varies depending on the application of the filter. The sediment obtained from bleaching earth filtering is either a waste, or the fat residue is recovered from it. Some sediments after filtration form a waste which is incinerated.

Centrifuging is widely used for clarification of oils. Centrifuges are particularly suited for the removal of fine suspensions, which cannot be removed by sedimentation. The volume of the sludge collector must be larger when the content of the suspension is higher and water is present. Such centrifuges, called "purifying", operate in a continuous mode removing gums and carrying away the refined oil. "Clarifying" centrifuges are used for prepurification of oil before refining, for clarification of salad oils and for purification of fish oils. Finally, "separating" centrifuges equipped with helical conveyors for continuous discharge yield a totally clear oil, and the maintenance is minimal. Miscella is usually purified by filtration or by means of hydrocyclones, while pressed oil is first left for sedimentation and then centrifuged in order to remove even the finest impurities. Paper filters are also used, but they are more cumbersome.

1.1.7.2 Dewaxing

Waxes, as esters of fatty alcohols and fatty acids of high boiling points, are slightly soluble in oils. They occur mainly in sunflower, maize and linseed oils and cause their turbidity at lower temperatures.

Sunflower oil wax is a mixture of esters of fatty acids, mainly C_{20} (44%) and higher — up to C_{28} (in smaller amounts), and C_{22}—C_{30} alcohols. Physical properties are similar to that of carnauba wax. Depending on the temperature, these waxes are partly soluble in oil.

The wax content in crude oils varies from a few mg/kg to 2000 mg/kg (Haraldsson, 1983). In order to make the oil transparent at lower temperatures it is necessary to reduce the wax content to ca. 10 mg/kg.

The traditional dewaxing method consists in a slow cooling of oil, crystalization of the waxes and their separation by refining. These operations are carried out either after bleaching, or after deodorization. Due to the necessity of obtaining crystals suitable for filtration, it is crucial that the wax is cooled slowly. A modern dewaxing method is illustrated in Fig. 1.21. The oil is cooled first by the outflowing dewaxed oil, then in a heat exchanger and in a crystallizer to 6–8°C. Auxiliary agents are usually added in an amount dependent on the wax content. Crystallization time is equal to at least 4 h, while maturing time is at least 6 h. The oil is subsequently cautiously heated to 18°C and filtered in a horizontal filter press with a rate of 40–50 $kg/m^2 \cdot h$. The system described is suitable for oils of wax content lower than 500 mg/kg. However, new sunflower varieties contain up to 1500 mg/kg waxes and the methods used till now are unsuitable

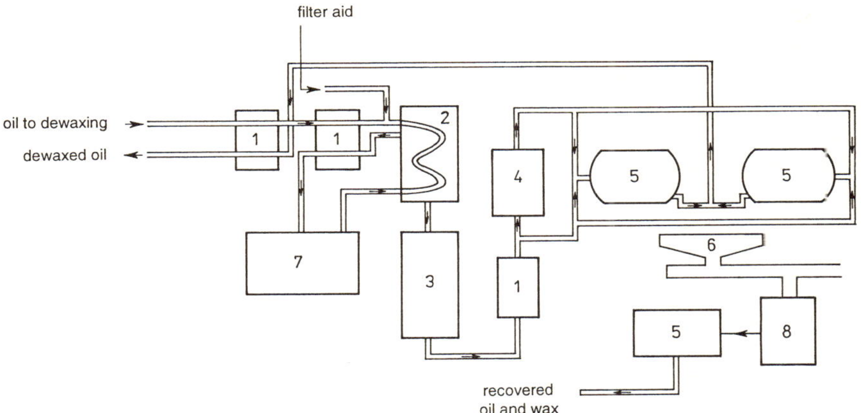

Fig. 1.21 Dewaxing of sunflower oil (Haraldsson, 1983): 1—heat exchanger, 2—crystallizer, 3—maturing apparatus, 4—filter supplying tank, 5—filter, 6—sediment tank, 7—cooler, 8—flow-through tank.

since the filtration takes a long time, oil losses are greater and the necessary auxiliary agents are too expensive. In order to avoid these difficulties, dewaxing is carried out in the crude oils. For this purpose the oil after solvent extraction is cooled down to 25°C and maintained at this temperature for ca. 24 hours. It is next hydrated at the same temperature, matured for 30 min and then centrifuged. Such a process reduces the wax content to 200–400 mg/kg.

A new dewaxing method combined with alkali neutralization has been recently developed. Waxes are separated from the oil in an aqueous suspension using a centrifuge (Fig. 1.22). Crude oil of ambient temperature is heated by heat exchange with the neutralized oil to 80°C, then to the

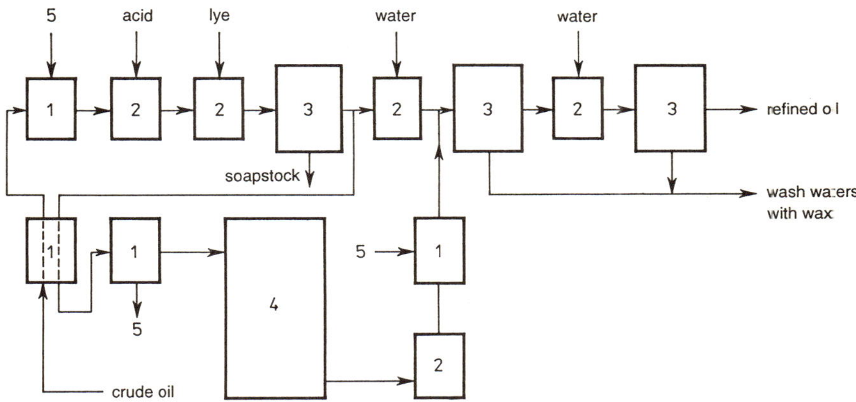

Fig. 1.22 Dewaxing with simultaneous alkali neutralization (Haraldsson, 1983): 1—heat exchanger, 2—mixer, 3—centrifuge, 4—maturing apparatus, 5—heat carrier.

neutralization temperature, and degummed with phosphoric acid. Next it is neutralized with sodium hydroxide and separated. After cooling in a heat exchanger to 30°C and 8°C it is stored for 4–5 h and slowly stirred. Then 4–6% of water is added and the whole is heated to 18°C with lukewarm water. An aqueous soap phase is formed during stirring. This phase moistens and extracts small wax crystals, forming a suspension. The heavy phase formed is separated from the oil by centrifuging. The oil is again mixed with a small amount of water and centrifuged in order to totally remove the waxes. This method is less expensive, since the cost of the auxiliary agent is avoided, and losses of oil and labour costs are reduced. The wax suspension is mixed with soapstock prior to its splitting. However, this waste can also be treated separately in order to utilize the waxes and the oil for further refining (Haraldsson, 1983).

Waxes and hydrocarbons contained in sunflower seed can be removed by washing the seed with boiling hexane (Morrison, 1982). About 92–95% of them are removed in this way. The wax material removed contains 70–79% of waxes and 13–30% of hydrocarbons. Dewaxing can also be accomplished after refining of the oil. In this case also an aqueous solution containing wetting agents is used. The suspension of waxes in the aqueous phase is centrifuged, thus separating the waxes from the wetting agent, which is recycled to the process (Fig. 1.23). Dewaxing in miscella is also possible (Cavanagh, 1988).

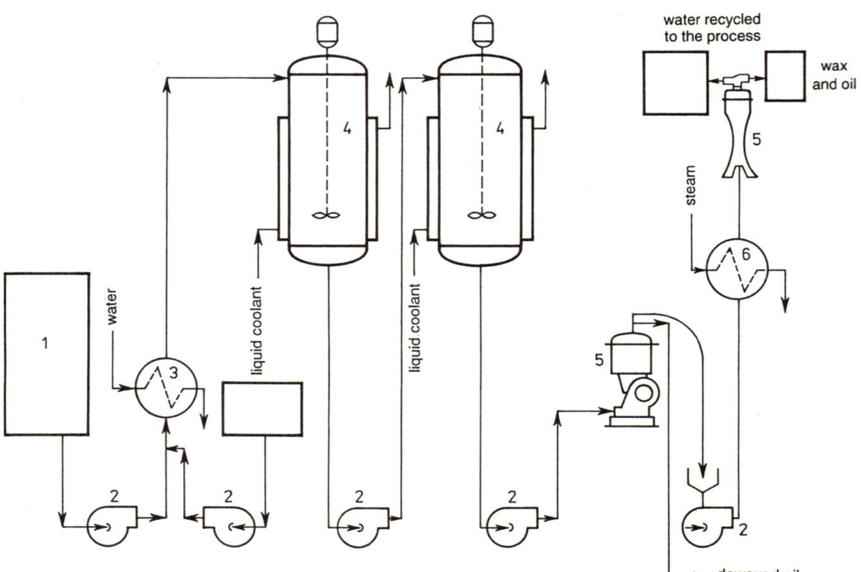

Fig. 1.23 Dewaxing of refined oil (Latondress, 1983): 1—refined oil container, 2—pump, 3—water cooler, 4—crystallizer, 5—centrifuge, 6—heater.

1.1.7.3 Alkali neutralization of oils

The traditional batch neutralization method, still used in small refineries (up to 10 t/d), apart from certain advantages has serious drawbacks, e.g. formation of large amounts of acid waters from soapstock splitting and washing, which is dangerous for the environment. Introduction of the Zenith system constituted an advance in the refining process. In this system the oil is treated with phosphoric acid in order to remove the emulsifiers, and next is introduced to the bottom of an apparatus filled with diluted sodium hydroxide. The oil, having lower density, rises in the form of tiny droplets to the upper part of the apparatus, where it is accumulated as neutralized oil containing little soapstock. As the process proceeds, the hydroxide becomes more and more diluted and contains increasing amounts of soapstock. Therefore it is replaced with a new solution, and the spent solution is treated with sulphuric acid in order to split the soaps and recover fats. Simultaneously with this method the application of centrifuges for the separation of gums and soapstock, as well as washing of oil, found widespread application.

In recent years the scope of research on further improvement of alkali neutralization has broadened, taking into account among others a reduction of the environment pollution. The environment is endangered by soapstock and wash waters formed during washing of oil.

Soapstock contains 10–40% of fatty substances (calculated as fatty acids). Its amount is equal to 3–20% of that of the crude oil. Therefore utilization of soapstock is of great economic importance. Moreover, any unused fatty matter increases the contamination of wastes. When the refinery is situated in the same plant as the solvent extraction unit, then the crude soapstock from neutralization can be recycled to meal in the desolventizer-toaster. When the total content of fatty acids exceeds 30%, the soapstock solidifies easily, which makes it necessary to use heated tanks and pipes and to maintain a temperature of 60°C (Woerfel, 1983). During storage, soapstock stratifies, forming a bottom aqueous layer. On heating to boiling it tends to foam, particularly when the oil from which it is obtained had not been hydrated.

The batch method is still often used for soapstock splitting. The process is carried out in acid-proof reactors. Wooden tanks lined with an acid-resistant material are also used. In the traditional splitting method the soapstock is boiled with an excess of sulphuric acid of 80–95% concentration for 2–4 h. The mixture is next left for at least 4–6 h in order to enable phase separation, and the lower aqueous phase, called acid water, is removed. The upper layer is washed with 25–50% of water, boiled and left for phase separation to occur. After removing the lower layer, the upper one can be stored or shipped. The lower layer has low pH and high BOD. However, if the pH is adjusted to a proper level and the immiscible components are separated (see Chapter 4), then it can be subjected to biodegradation.

The composition and quality of soapstock depend on numerous factors. Soapstock formed during alkali refining by a continuous method is usually better than that obtained by the batch method, since it contains a smaller amount of neutral fat. Moreover, the continuous method reduces the losses due to emulsification and entrainment of the droplets of neutral fat. The losses are thus reduced by 20–30%. Usually the soapstock from oils of higher AV contains a smaller amount of neutral fat compared to the soapstock obtained from oils of low FFA content. On the other hand, when the content of FFA in crude oil is excessively high, i.e. 8–20%, the losses significantly increase due to emulsification and entrainment (Wong, 1983). Degumming decreases the content of FFA in soapstock. Apart from the amount and concentraction of the alkali, other factors also significantly influence the characteristics of soapstock, viz. the manner of introduction of the alkali, contact time, the manner of stirring, temperature, etc.

There are a few technologies of continuous soapstock processing. They have been described by Braae (1976), Crauer (1970) and Morren (1981).

After separation by centrifuge the soapstock is blended with waters from oil washing and the mixture is subjected to splitting. The material is pumped in defined portions to a stirring tank, in which it is treated with sulphuric acid and heated. The content of the stirring tank flows gravitationally to a separator, where phase separation takes place. The lower aqueous layer, formed by acid water of a very low pH, is neutralized with sodium hydroxide and disposed of to the sewage system. The upper layer, containing fatty acids, is sold.

Methods based on application of centrifuges for the separation of the aqueous phase from the oil phase are expensive. Instead, a skimming tank of such a volume that the contents can stay inside for 2–3 h can be used. Good phase separation takes place during this time. The aqueous part is neutralized to pH 5.5–6.0 and disposed of to wastes. The oily phase is washed with 5–10% of hot water in order to remove the residue of the mineral acid. Fatty acids prepared in such a manner can be stored or shipped in tanks made of mild steel.

The main disadvantages of traditional soapstock processing are not only the corrosion of the equipment, but also inaccurate separation of fat from the acid water, which results in losses of fatty matter and strong contamination of the wastes. Special resistant materials must be used due to the need for the anti-corrosive protection, and that increases the investment costs of the equipment and the buildings. To avoid this, a continuous process has been proposed, not requiring large equipment, but still assuring a reduction of the level of fat in the acid water down to 10–28 mg/kg in the case of mixed rapeseed and soyabean soapstock. For this purpose the coalescence of fat that has not been separated in the preceding gravity separation is included. A glass fiber bed has been used as the coalescence device (Mag *et al.*, 1983). After flowing through this layer, another gravity separation is carried out, after which the neutralized water can be directly disposed of to the sewage system. Pilot plant experiments revealed that the bed must be regenerated ca. every 65 h of operation. The duration of the operating period depends on the bed thickness. Regeneration time is ca. 1 h. Acid water freed from fat in this way is neutralized and disposed of to the sewage system (Fig. 1.24).

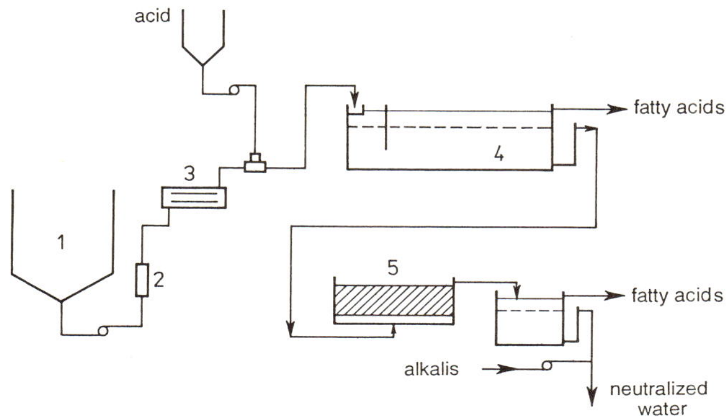

Fig. 1.24 Continuous soapstock hydrolysis (Mag *et al.*, 1983): 1—soapstock tank, 2—flow meter, 3—heat exchanger, 4—skimming tank, 5—glass-fiber bed.

Although the methods described significantly reduce the risk of contaminating the wastes with acids and high BOD, yet there is always the necessity of installing expensive devices requiring high operating costs. For this reason a method for soapstock drying has been developed (Beal, 1981). It is based on drying soapstock acidified with 25% sulphuric acid under a reduced pressure, using a film-type evaporator or an evaporator with natural circulation for this purpose. The resulting product is waxy at room temperature and contains 54–71% of fatty matter. The only waste is water. If the soapstock originates from a non-hydrated oil, then the product contains large amounts of carotene and xanthophyll (Fig. 1.25).

According to Canadian sources, splitting of soapstock from 00 rapeseed oil, neutralized in a batch manner, is performed in the following way: after 10–15 min of stirring of neutralized oil, the stirrer speed is usually reduced from 40 to 8 rpm, and if the neutralization is carried out at a temperature lower than 70°C, it is heated to this level in order to facilitate the agglomeration of soapstock particles and partial melting of soaps, which facilitates liberation of the oil. After stopping the stirrer the mixture is left for at least 1/2 h in order to enable a good phase separation. The soap layer is removed from the bottom, and the upper layer is washed at least four times with 5–10% of hot water. After the last washing ca. 200 mg/kg of citric or phosphoric acid is added in order to split the last traces of soaps. The oil is dried at 105°C with vigorous stirring (Teasdale and Mag, 1983).

Wash waters contain fatty substances, up to 75–85% of which (calculated as fatty acids) can be recovered in the following way. A small amount of

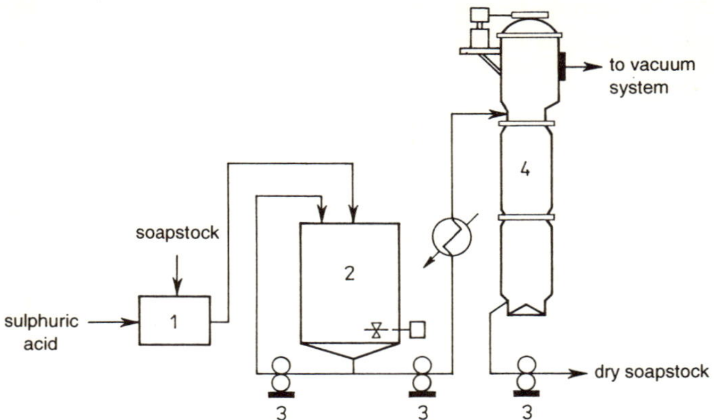

Fig. 1.25 Soapstock drying (Woerfel, 1983): 1—mixer, 2—flow-through tank, 3—pump, 4—vacuum turbo drier.

lignin is added to the hot wash waters from alkali neutralization of soyabean oil, and they are subsequently decanted. Next the pH is adjusted to ca. 2. This method reduces the contamination of this waste by 90–95% (Poley, 1957). An improvement can be made by introducing countercurrent washing of the oil with water. This system requires only 4% of water, hence the amount of acid wash waters, being a cumbersome waste from alkali neutralization, is reduced (Haraldsson, 1983).

Post-refining fatty acids are the most important by-products formed during the neutralization of oil. They are obtained either from soapstock (when alkali neutralization is applied), or as a condensate when distillation is used. In the former case the soapstock is isolated from the neutral oil with a centrifuge. Water is added in order to facilitate pumping when the viscosity of the soapstock is too high. Then it is hydrolyzed and thus the technical post-refining fatty acids are obtained, called usually "acid oil".

An improved method of processing has been patented (Red and Ilagen, 1978). The method enables over 90% recovery of free fatty acids from crude soapstock and good separation of the FFA from acid waters. For this purpose the crude soapstock with an addition of 20% of sodium hydroxide solution of a density 1.222 and 10% of a 17% sodium sulphate solution is heated to ca. 80–95°C for 2 h. The completely saponified soapstock is acidified for 2 h at 82°C and separated in a centrifuge.

In many countries fatty acids from soapstock are distilled. When the soapstock is obtained from batch neutralization, it can contain as much as 40–50% of neutral oil. In this case the fatty acids are hydrolized with water in an autoclave, partly yielding glycerin and reducing the losses during the distillation. On the other hand, when neutralization is accomplished using separators, then the soapstock contains only half of the oil occurring in the batch method. In such a case it is better to perform the final saponification at 150°C. Under such conditions the phospholipids are also saponified, which significantly facilitates the splitting and obtaining of 98% acids by distillation. Losses are negligible.

Neutralization with ammonia. Neutralization of oils with sodium hydroxide and subsequent splitting of the soapstock causes the formation of large amounts of sodium sulphate, which is disposed of with the wastes. For instance, in 1973 in West Germany the wastes from an edible oil refinery contained 4468 t of sodium sulphate (Pardun, 1979). Because of such a heavy environmental pollution the distillation method of neutralization is gaining popularity. Formation of non-volatile oxidation products and polymers constitute the disadvantages of this method. Therefore already in 1959 it was proposed to neutralize fatty acids with ammonia. This method has not found widespread application since ammonia does not effect decolorization to such an extent as sodium hydroxide. Recently, however, the method again becomes interesting due to the still more rigorous limitations concerning the composition of wastes, since the ammonia is recycled and does not feed the wastes with salts. On the other hand, the investment costs are higher in this case compared to the classical alkali neutralization. The process has the following course. The oil is hydrated in order to remove lecithin. The degummed oil is treated at 30–70°C with 150% of the calculated amount of ammonia, and the resultant soapstock is separated in a separator. The soapstock is evaporated, and the evolved ammonia is recycled to the neutralization. Crude fatty acids containing 20–30% of neutral oil are obtained as the by-product. If necessary, the neutralized oil is treated at 95°C with 2% of a 0.5 M/l sodium hydroxide solution, and is subsequently washed and bleached. If the oil has not been hydrated, then it is degummed by an addition of 2% of a 5% citric or formic acid solution and neutralized with ammonia after the isolation of gums. The soapstock is evaporated in order to recover the ammonia. Lecithin enriched in fatty acids is obtained as a by-product (Pardun, 1979).

1.1.7.4 Bleaching of oils

Bleaching aims not only at the removal of pigments, but also of soaps, gums, sulphur compounds, metals (e.g. iron) and oxidized compounds from the decomposition of peroxides. According to Ong (1983), bleaching earth also adsorbs polymers, mainly polar oxypolymers. Bleaching earth removes phosphorus to the level of 1 mg/kg provided that the oil has been previously degummed with acid. Thorough removal of the earth after bleaching is very important, since already a 0.01% content significantly reduces the oil stability.

Drying of seed by direct contact with exhaust gases causes the edible oil to contain polyaromatic hydrocarbons. In some crude oils, e.g. sunflower and coconut, the occurrence of 13 polyaromatic hydrocarbons has been established. Depending on the volatility, these compounds can be divided into two groups. Some of them are volatile under the conditions of usual refining, hence adsorbtion on active carbon can be applied. It is accomplished after the traditional bleaching with bleaching earth. The carbon is added 5 to 10 minutes after the introduction of the earth in an amount dependent on the degree of contamination of the oil with the hydrocarbons. For example, 0.4% of carbon is used for coconut oil containing 3000–4000 mg/kg of PAH, while 0.25% is sufficient in the case of sunflower oil containing 300–400 mg/kg of PAH. Of great significance is the adsorptive capacity of the carbon towards these compounds. Wendt (1981) found out that only certain kinds of active carbon are suitable for the process described. There is no data available whether rapeseed dried with exhaust gases also contains PAHs.

Physical refining removes odours, pesticides and FFA, and transforms carotenoids. Other side-products must be removed during the initial refining of the oil. The following substances remain in oil after the traditional hydration: carotenoids, chlorophylls, other pigments, part of phospholipids, metals like iron, copper, calcium and magnesium, free carbohydrates, glycolipids and oxidized lipids. Removal of these compounds depends primarily on the properties of the bleaching earth used as the adsorbent and on the technological conditions (Segers, 1983). The most important properties of the earth are the adsorptive capacity, activity and ion exchange capacity.

— Adsorptive capacity depends on the surface area, as well as pore size and number. The presence of soaps and phospholipids significantly reduces the adsorptive capacity, e.g. towards pigments.
— Catalytic activity depends on the action of the acid during the activation of the earth and on the bleaching temperature.
— Ion exchange capacity. The majority of bleaching earths consist of silicates with aluminium ions. These ions are partly removed by weathering and during activation, and are replaced with protons. In this way the earth becomes able to bind metals.

The amount of bleaching earth is equal to 0.5–2.5% with respect to the oil. It contains 20–40% of oil, which is very valuable and is usually recovered. However, a too intensive recovery can cause a partial desorption of the impurities and thus a decrease of the value of the oil.

The amount of the absorbed oil depends on the activity and particle size of the earth.

Processing of bleaching earth. The suspension of bleaching earth in oil is pumped at a controlled rate to one of the filter presses. When it becomes filled up, it is disconected and blown with compressed air to remove the oil. Then steam is additionally forced through. The content of the oil in the bleaching earth depends on the kind of the adsorbent and the filter, the filtering conditions, the manner and the extent of oil removal from the earth. Usually the amount of the absorbed oil is equal to 25% (calculated per mass of the bleaching earth) after blowing with steam and to 20% after rinsing with the circulating water (Ong and Sinkeldam, 1983).

Recovery of the oil can be carried out either in the filter press, or after removing the filter cake. Depending on the method used the composition and the amount of the waste is different, and that can determine how it is used. The simplest method of recovering of the absorbed oil is to boil the earth with diluted sodium hydroxide solution of the density 1.02. Under such conditions the oil is displaced without saponification. Some salt is subsequently added in order to break the emulsion and hot water is introduced. After phase separation the oil layer is collected. The quality of the oil depends on the contact of the filter cakes with air, nevertheless such an oil is usually mixed with soapstock.

Extraction of the oil in a filter press can be acomplished either with hot water, or with hexane. In the former case water at 95°C is forced at a pressure of 0.5 MPa through the cake for ca. 30 min. A reduction of the fat content to 20%, i.e. by 55–70%, is achieved. The ratio of the amount of water to the amount of the bleaching earth is between 5:1 and 20:1, usually 10:1. The cakes are subsequently blown with steam, recovering 90% of the oil within 10 min. Next the cakes are dried by forcing air through. This method cannot be applied for fish and linseed oils (Patterson, 1976).

In the latter case the cakes are extracted with semi-miscella, containing 3% of oil, for 20 min, yielding a 12% miscella, which is subsequently distilled. Next pure hexane is passed through the cake which is then blown with steam. The initial content of the oil in the bleaching earth equal to 30% is reduced down to 5%. Since the extracted earth is dusty, it is sprinkled with water and transported to a waste dump (Ong, 1983).

Hexane extraction is profitable only in the case of expensive and stable oils, like peanut, olive oil and cocoa butter. This is due to the high cost of this process and to the fact that oils of high IV absorbed by the earth are very unstable. Due to high investment and labour costs, as well as the hazard related to the use of inflammable solvent, this method is seldom used in industry.

Due to the increasing high labour costs connected with emptying and cleaning the traditional filter presses, they are gradually replaced by self-cleaning closed filters, automatically controlled.

Another method yielding a good quality oil from the extraction of the cakes with hexane, with an additional advantage of a simple utilization of the deoiled earth, consists in replacing the fat with water. A column presented in Fig. 1.26 is used for this purpose. The cakes containing the earth are first disintegrated in a tank equipped with a stirrer, and next they are mixed with the solvent, forming a fine suspension. This suspension is continuously fed to the extraction zone of the column. Its diameter is calculated in such a way that the rate of upward miscella flow is lower than the rate of sedimentation of the earth in the miscella. Separation of the earth from the miscella is thus achieved (Weber, 1983).

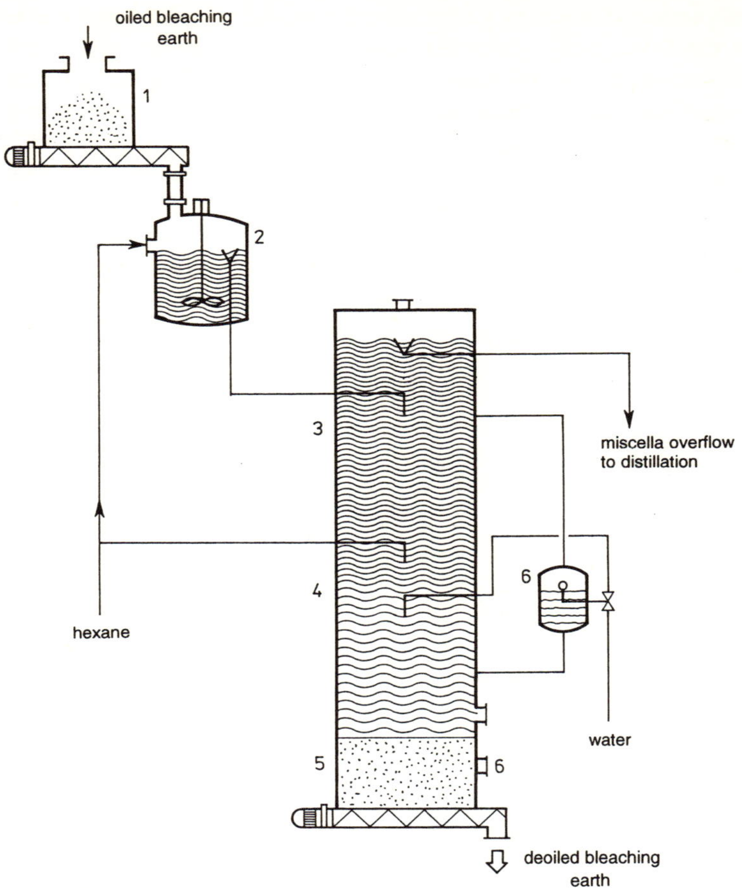

Fig. 1.26 Apparatus for hexane extraction of oil from bleaching earth in a three-phase system (Weber, 1981): 1—earth container, 2—apparatus for mixing the earth with the solvent, 3—extraction zone, 4—separation zone, 5—decantation zone, 6—level controller.

The column is filled 1/3 with water and 2/3 with hexane. The mixture from the stirrer feeds the column in 2/3 of its height. Fresh hexane is supplied above the water/hexane interface. Since the miscella moves upwards, a countercurrent extraction takes place. The miscella freed from the earth leaves the column at the top through an overflow and is directed to the distillation and regeneration equipment. Hot water at the bottom of the column replaces the hexane adsorbed by the earth. The level of water is automatically controlled. The deoiled and hexane-free earth settles at the bottom of the column. After reaching a certain height this layer becomes impermeable for water and can be removed with a helical conveyor. In this method 1% of the oil is left in the earth and the miscella formed has a concentraction of 8–10%. The water temperature can be slightly higher than the boiling point of hexane. The oil recovered in this way has lower AV and PV and is lighter in colour than the crude (soyabean) oil. No wastes are formed, since the condensate of steam used for expelling hexane is supplied to the lower part of the extraction column in order to supplement the water carried away by the earth leaving the column. This earth has an almost constant consistency, can be easily transported and can serve as a filling material instead of sand.

Used bleaching earth has been de-oiled using liquid carbon dioxide at high pressure (Waldmann & Eggers, 1991). Typical conditions were 500 bar pressure at 80°C, giving a recovery of about 97% of the adsorbed oil. Extracted palm oil and rapeseed oil were unchanged in fatty acid composition, and the oil quality was judged to be suitable for processing into an edible grade. The activity of the de-oiled bleaching earth had diminished by 50% or more.

The regeneration of spent bleaching earth has been achieved by a 'Zimmerman' type wet oxidation (Kalam & Joshi, 1988). The earth is first de-oiled and dried. It is then suspended in water at a 5% concentration and treated at 200°C and an oxygen partial pressure of 0.5 MPa in a stirred autoclave. The treated earth was tested on ground nut and cottonseed oils and had the same decolourizing capacity as unused earth over at least four cycles of regeneration.

Neutral oil loss in bleaching is significantly reduced by the use of amorphous synthetic silica adsorbents in place of bleaching clays (Welsh, 1990). Silica effectively removes all impurities except chlorophyll. Where this is present the use of silica in conjuction with bleaching clay is effective. Silica also removes traces of soap without loss of activity.

1.1.7.5 Deodorization of oils

Rational handling of the degumming, neutralization and bleaching wastes does not usually require new investments or major changes in the technology of these processes. What it requires is only an agreement with the buyers on the possible changes in the quality and the price based on a more economical utilization of the wastes in the feed or chemical industry. The problems therefore are rather of an organizational, and not technological character.

Utilization of wastes from deodorization requires a change in the vacuum system and in the manner of collection of the valuable products, yet it gains the possibility of yielding new by-products of high price. These products can be further refined in the chemical industry. Additional benefits consist in a substantial reduction of the contamination of the wastes, hence of the fines paid for surpassing the limits. More extensive changes in the equipment allow reducing the atmospheric pollution with noxious fumes.

The deodorization condensate consists of FFA, neutral fats and UM. Upon contact with the barometric condenser waters the free fatty acids react with calcium carbonate and form calcium soaps. The distillate consists of fine flocks flowing to the surface and coagulating during the settlement.

The oil loss during deodorization in a modern continuous equipment is equal on average to 0.2–0.3%, 25 to 45% of which are the free fatty acids. The magnitude of the losses (and their composition) depend on the kind of oil — they are smaller for high-erucic rapeseed oil than for soyabean oil, and particularly for palm oil (Brekke, 1980 b), provided that a suitable negative pressure is maintained. For example, 2.5 t of the condensate is obtained during processing of 1000 t of soyabean oil. After substracting the FFA, ca. 1.6 t of lipid distillate remains. Assuming that it contains 10% tocopherols, a raw material for obtaining 160 kg of these substances is obtained. However, significant losses during extraction in the pharmaceutical industry must be taken into account due to the instability of these compounds.

The condensate is usually directed to a trap, where it flows slowly in order to allow the fatty matter to accumulate at the surface, from where it can be collected. The separated components are heated in order to remove water, and often combined with soapstock, since the high content of unsaponifiable matter and metallic soaps, as well as the unpleasant smell, render any other utilization of the condensate impossible.

Water solubility of the main fatty acids of coconut oil and palm kernel oil is equal to:

Acid	mg/kg
caprylic	1100
capric	100
lauric	12
myristic	6

These values limit the possibility of removing the fatty substance contained in wastes by physical methods. Similarly their volatility hinders their condensation in the scrubber placed at the gas outlet from the deodorizer. This is evidenced by the fat content of water leaving the barometric condenser, e.g. (Young, 1983):

Deodorized oil	mg/kg
rapeseed	10
soyabean	10
palm	25
coconut & palm kernel	50

Pesticides used in soyabean growing are gradually removed during the subsequent processing stages. It has been established that dieldrin behaves similarly to aldrin. Deodorization constitutes the most effective method of their removal. Pesticides have been found in hulls, soapstock, oiled bleaching earth and deodorizer scum. Large differences in the content of aldrin and dieldrin have been established (Chaudry *et al.*, 1976).

Development in equipment. Till the end of the fifties the part of the deodorizer scum accumulated in the barometric well was utilized, while the rest was disposed to wastes, which was dangerous for the environment. A modified system of distillate recovery is nowadays increasingly often used. The fumes leaving the deodorizer are first cooled down to a temperature sufficient for the condensation of the higher boiling components, but not of the steam used for the deodorization. This operation takes place in a cooling scrubber mounted between the outlet of the deodorizer and the booster. The cooled condensate constitutes the cooling medium. The condensate collected at the bottom of the scrubber is removed through a heat exchanger and recycled to the scrubber. An automatic device controls the level of the liquid in the tank, and the excess is removed as a by-product. The fumes leaving the scrubber pass through a mist removal section in order to stop the entrained droplets. In order to remove the residues of fats an additional cooling tower is mounted between the booster and the first condenser. In this tower the distillate is cooled down to 60°C at a pressure of 5–8 kPa. The degree of its recovery is estimated at 95%. Measurements carried out on an industrial scale on soyabean oil containing 0.12% of tocopherols revealed that 0.35–0.45% of a distillate containing 10–14% of tocopherols is obtained. The parameters of this process were the following: deodorization temperature 274°C, 3–5% of live steam, 0.7–0.9 kPa pressure. The deodorization time in a semi-continuous Votator deodorizer is 15 min, while in the continuous EMI deodorizer it is 2 min. A reduction of the amount of live steam or a decrease in temperature results in a reduction of the amount of the distillate and its content of the tocopherols.

The old method of obtaining the deodorizer scum comprised a single flow of water used for the condensation of vapours of organic substances removed from the deodorizer. This water has been treated as a factory waste. The statutory limitation of the amount of wastes resulted in the introduction of closed circulation systems of cooling water. In these systems the water from the vacuum device is cooled in a cooling tower and recycled. However, an organic deposit settles in the tower with time. This deposit must be removed after stopping production. Moreover, when this deposit decays it causes an unpleasant smell in the vicinity of the tower and the surroundings. Introduction of a modified cooling tower without the packing, where the water was sprayed, constituted a partial, yet expensive solution to this problem. Cleaning of the spray nozzles also required much labour, and the smell had not been neutralized. Further improvement led to a solution presented in Fig. 1.27. The fumes from the deodorizer together with the steam from the booster flow into a closed tower, where they are cooled with the recycled condensate. The condensate collected at the bottom is passed through a heat exchanger and recycled to the spray in the tower. An automatic device controls the level of the condensate in the tank and passes the excess of the condensate to storage. With this system valuable by-products are recovered, viz. sterols and tocopherols. The vapours leaving the distillate recovery tower are condensed using water. In order to close

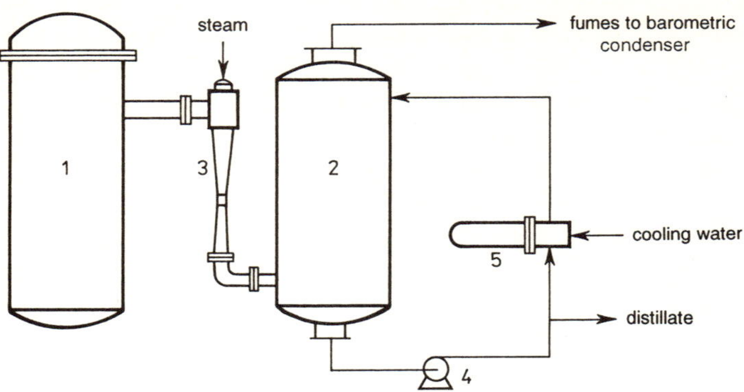

Fig. 1.27 Obtaining of deodorization scum (Gavin, 1978): 1—deodorizer, 2—distillate tower, 3—vacuum booster, 4—pump, 5—cooler.

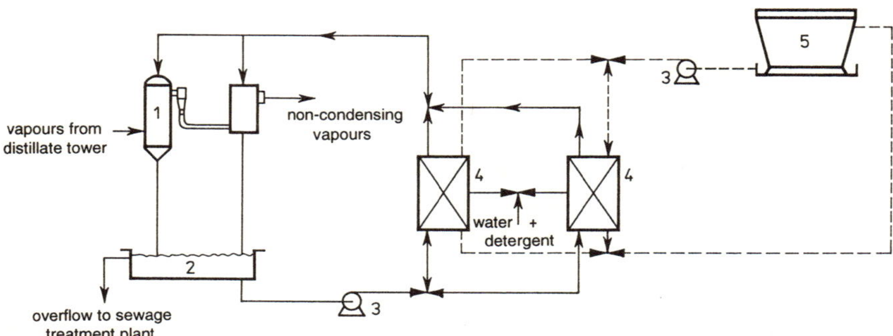

Fig. 1.28 Circulation of water from the deodorizer condenser (Gavin, 1978): 1—barometric condenser, 2—hotwell, 3—pump, 4—heat exchanger, 5—cooling tower.

its circulation a system presented in Fig. 1.28 is used. Pure cold water for supplying the tower is regenerated in the following way. The contaminated aqueous layer from the condensate trap below the condenser is cooled in a heat exchanger with cold water from the cooling tower before its introduction to the condenser. As there is no contact between these two water streams, the cooling tower does not become contaminated. This enables the application of packed towers. Organic substances are trapped in the condensate tank, from where they are fed to sewage-treatment plant (Gavin, 1978). A second heat exchanger is applied in order to assure a proper temperature of the cooling water. In due time, or when the temperature starts to increase, the system is switched to the second exchanger. The first one is then cleaned with water, containing a surfactant. The equipment operates automatically.

Low-boiling fatty acids are not condensed in the condenser and are liberated to the atmosphere. Therefore in some installations the outlet of the gases is introduced to the barometric well below the level of the con-

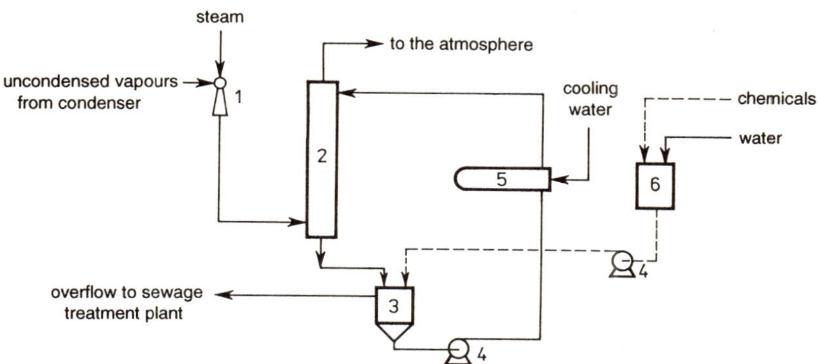

Fig. 1.29 Vapour scrubbing system (Gavin, 1978): 1—vacuum jet, 2—scrubbing tower, 3—scrubber tank, 4—pump, 5—cooler, 6—reagent tank.

densate. The vapours are thus dissolved in the aqueous phase and if there is no water recycling system they enter the cooling tower. On the other hand, if the water circulation in the system is closed, the vapours are concentrated in the well and cause an unpleasant smell in the surroundings. A device schematically depicted in Fig. 1.29 is used to avoid this inconvenience. The packing of the scrubber assures a sufficient contact time and area to absorb the volatile substances. A condensed water vapour from the ejector is collected at the bottom of the scrubber. This condensate also absorbs a certain amount of volatile substances. Their residue is again directed to the scrubber. The entire liquid phase is passed through a cooler (heat exchanger) and used for spraying in the scrubber. Sodium hypochlorite, calcium hypochlorite or potassium permanganate are used as auxiliary agents in odour control. These compounds are introduced to the closed system.

The amount of wastes from deodorization, distillatory method of neutralization and distillation of fatty acids can be significantly reduced by an application of the vacuum system proposed by Stage (1982). Figure 1.30 illustrates the application of a barometric condenser sprayed with organics-free water circulating through a heat exchanger. The residue of the organics is lower than 20 mg/kg. A cooler, removing liquid fatty acids is situated before the condenser. A proper choice of the temperature of water supplied to the condenser enables a condensate of the organic substances to be obtained in liquid form. Using such a solution the amount of wastes containing organic substances is equal to 35 kg/t in the case of distillatory neutralization or 65 kg/t for fatty acids distillation. In order to remove the fatty acids easily it is necessary to maintain proper temperatures in the cooler, after the ejector and in the separator.

A reduction of the amount of lipids in the condensate can be achieved in two ways. The first is separation before the barometric condenser, while the second is separation by centrifuging or other methods of breaking the emulsion after concentration. Such a concentration takes place when the amount of water used in the condenser is reduced. A scrubber can also be

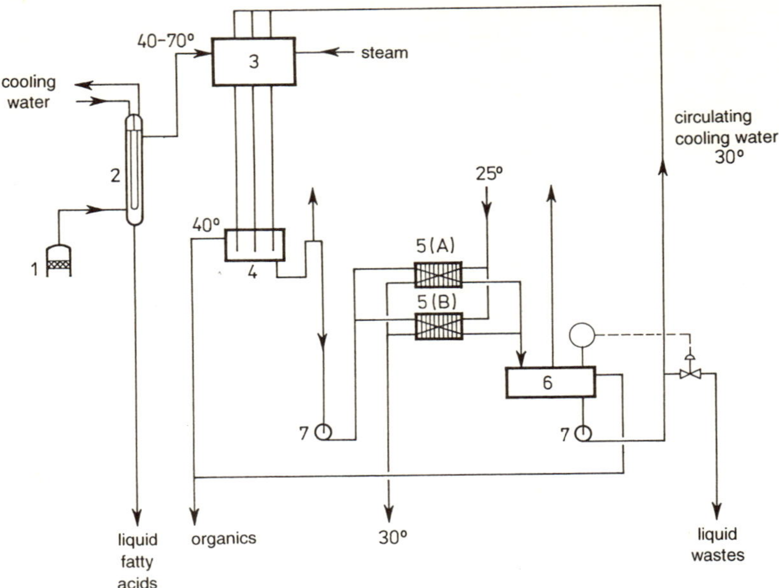

Fig. 1.30 Deodorizer vacuum system (Stage, 1982): 1—condenser, 2—cooling drip tube, 3—multi-stage ejector, 4—condensate tank, 5—plate heat exchanger, 6—separator, 7—pump.

placed at the pressure side of the booster and water can be used as the washing medium. Cooling of the vapour from deodorization results from the evaporation of water used for spraying. Another method consists in an application of two condensers before the booster. In the first one the vapours from deodorization are cooled down to 50°C. Under such conditions most of the lipids are separated in a liquid form. The vapours are subsequently cooled down to 15°C, and the rest of the lipids is separated in solid form. This section must therefore be periodically heated in order to liquefy the solid lipid fraction. Lipids separated in such a way are collected in a separate receiver (Kroll, 1980).

The condenser system in the equipment used for deodorization has been recently improved with respect to steam and electric energy savings (Schumacher, 1983).

In order to decrease the content of fatty matter in the condensates from deodorization and physical neutralization, and simultaneously to obtain a condensate of UM content lower than 30%, a vacuum system for two-stage condensation has been proposed (Liebing *et al.*, 1983). The pressure and the temperature at both the stages depend on the outlet pressure of the deodorizer and on the type of the oil being deodorized. For typical fats, viz. coconut, palm and sunflower, the pressure is 0.4 kPa, the deodorization temperature 240°C, and 1.2% w/w of steam is used. From the deodorizer the vapours flow to the collector, where at a pressure of 0.4 kPa and a temperature of 150–170°C the entrained oil droplets are trapped. From the

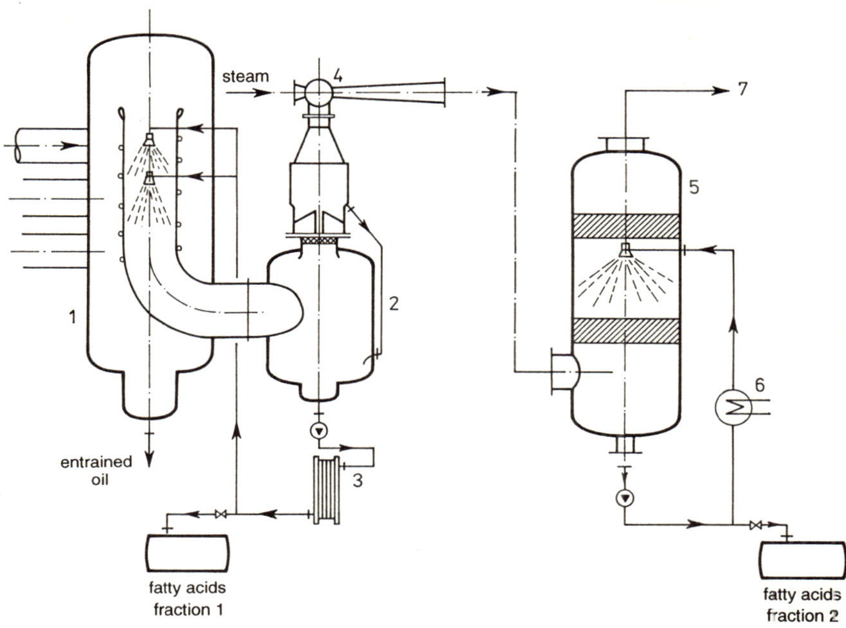

Fig. 1.31 Two-stage condensation of fatty acids in an apparatus for physical refining (Liebing *et al.*, 1983): 1—vapour collector, 2—first condenser, 3—plate cooler, 4—ejector, 5—second condenser, 6—cooler, 7—suction to the second condenser.

collector the vapours pass to the first condenser, where the pressure is 0.4 kPa and the temperature is 90–110°C. The first fatty acid fraction poor in UM is condensed. This fraction is partly used for the spray after cooling (Fig. 1.31). The extent of cooling is adjusted in such a manner that the circulating acids have a temperature higher by 2–3°C than their congeal point (titer). The excess of the acids is removed from the system. The vapours pass subsequently to the second condenser where the temperature is 40–50°C and the pressure is 1.9 kPa owing to the use of a second ejector. At the pressure side of the ejector the pressure is 7.5 kPa. The second fraction of the fatty acids is obtained in the condenser. This fraction is rich in tocopherols and sterols. A part of this condensate is recycled to the system after cooling. This system allows obtaining wastes containing 0.1–0.5 mg/kg of lipids, depending on the kind of oil. After this partial condensation the superheated water vapour contains only a small amount of fatty acid vapours. They pass to the vacuum devices and next to the barometric well. From the well the wastes pass through a cooler to the spray of the condensers. Pure water cooled in a cooling tower, hence not contacting the wastes, is used for feeding the cooler.

To remove the residues of the condensed lipids from the wastes, a second scrubber supplied with a saturated $CaCl_2$ solution at a temperature of –5°C can be used. Due to this, only vapours free of fatty substances enter the booster. These substances are passed after a repeated cooling through a

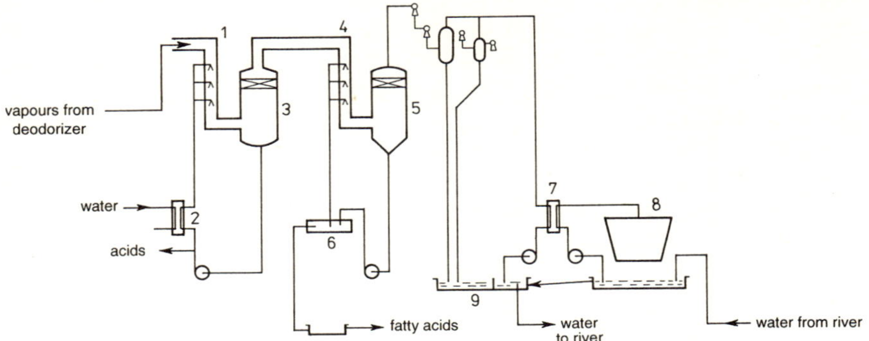

Fig. 1.32 Two-stage condensation of vapours from deodorization (Athanassiadis, 1982): 1—low-boiling acids condenser, 2—cooler, 3—fatty acids separator, 4—condenser supplied with brine, 5—condenser of acids boiling at higher temperature, 6—vacuum system, 7—plate cooler, 8—cooling tower, 9—condensate receiver.

skimming tank to further utilization (Fig. 1.32) (Athanassiadis, 1982). The wastes contain less than 8 mg/kg of fatty acids, which corresponds to 60 mg/kg COD.

Application of a surface condenser. The vapours from the deodorizer are passed to a vertical condenser, cooled with water circulating through a cooling tower. In order to assure a continuous operation it is better to use two small condensers in parallel (each of area e.g. 200 m^2) instead of one bigger condenser (of the area 400 m^2).

1.1.7.6 Losses during alkali refining

Removal of non-glyceride components from crude oils is always connected with certain losses. Crude oil usually contains 0.25–0.50% of water and 0.1–0.2% of gums and solid impurities in a suspended form. During their isolation neutral oil is lost in an amount equal to ca. 50–100% of the mass of these impurities. This oil can be partly recovered as a lower quality fat. The content of phospholipids in crude oil depends on its kind. It varies from < 0.1 to 2.5% A large per cent of fat remains in phospholipids during their isolation. Depending on the method of neutralization, the liberated free fatty acids, soaps or other compounds that can be relatively easily separated, can cause losses of the neutral oil to a varying extent. A quantitative measure of these losses can be among others the refining coefficient, being a ratio of the losses during the neutralization to the FFA content of the crude oil. The refining coefficient in batch alkali neutralization is usually equal to 1.3–2, while in the continuous method it is 1.3–1.6. However, it depends to a large extent also on the kind and the quality of the crude oil. After the separation of soapstock and its splitting the oil and aqueous phases are formed. The oil phase contains not only fatty acids, but also neutral oil, which is a refining loss. This oil is usually 10–50% of the total fatty matter of the oil phase. In an average batch refinery it varies between 25 and 30%. The next loss occurs during washing of the oil after neutralization. This loss is usually equal to 0.1% of the oil. Much greater losses take place during bleaching. They depend primarily on the amount of the adsorbent used, and they are usually much smaller when active carbon is used. However, bleaching earths are usually used. If for example 2% of bleaching earth is used, then 1.3–1.7% of fat is absorbed. Even when 1–1.4% of the oil is recovered, it is of poor quality and the loss still amounts to 0.3%. A further loss occurs due to an increase in acidity by 0.2–0.3%. This increase results from hydrolysis of the neutral oil due to the acidic activation of the bleaching earth. These FFA are next distilled off during the deodorization, but their quality is poor. Usually the loss of oil resulting from the deodorization is equal to ca. 0.1–0.2%, but part of the oil is entrained during the vacuum process, and a small fraction is hydrolized during deodorization. Together with FFA formed as a result of the action of the adsorbent the total losses during deodorization can amount up to 0.5%. An increase in the yield of refined oil requires increased investment costs and an improvement in the quality of the crude oil. Moreover, the losses depend on the chemical composition of the oil, hence on its kind. They will be different for instance for palm oil compared to rapeseed oil.

1.1.7.7 Physical neutralization

The amount of hazardous wastes can be reduced by physical neutralization. In this case the processes of alkali refining, washing, soapstock splitting and neutralization of acid water are unnecessary. A significant reduction of the amount and harmfulness of the wastes constitutes a primary advantage of this method. This is due to the fact that the wastes do not originate from soapstock processing and that they contain less of the entrained fat. Acidification of soapstock, which is most cumbersome with this respect, is eliminated. After the hydration of crude oil less than 0.03% of phospholipids remain, and their content is further reduced to 0.01% by an addition of 0.4–0.8% of bleaching earth. Owing to this, phosphoric acid does not have to be used. Soyabean oil prepared in such a manner can be directly neutralized by distillation and deodorized. The condensate obtained contains ca. 7% tocopherols.

The method aims at:

— a reduction of the FFA content from 5% to 0.03% or even less,
— obtaining a fully deodorized fat,
— performing the process in a manner not requiring higher costs than the traditional deodorization,
— a recovery of fatty acids removed by the steam.

Moreover, the equipment should be suitable for carrying out the traditional deodorization without deteriorating the quality or increasing the costs.

Instead of soapstock, only distilled fatty acids are formed in this process, which is crucial for the protection of the environment (Tandy and

McPherson, 1980). The method has been applied for refining palm oil, palm kernel oil, coconut oil, sunflower oil, maize oil, animal fats (Gavin, 1978) and low-erucic rapeseed oil (Daun, 1982; Roden, 1983). Losses during physical neutralization are lower than during alkali neutralization for oils of high FFA content. The neutralization and deodorization cycle in this method comprises three unit processes performed on a thoroughly degummed oil free from non-volatile by-products. In the first process the FFA are removed in a vacuum equipment by a multi-stage, countercurrent contact with steam. In the second process the pigments are transformed into colourless compounds by keeping the oil at an elevated temperature for a certain time, while in the third the oil is deodorized by a countercurrent contact with steam.

The oil is first heated to 80–90°C in order to remove air and water, the temperature is subsequently raised to 160°C and the oil is mixed for 10 min with concentrated phosphoric acid. Under such conditions the gums are precipitated and adsorbed by the added bleaching earth. Following the reduction of temperature to 80–90°C the earth is filtered off, and the oil is cooled down to 50°C.

The preliminary purified oil is continuously pumped through a filter and sprayed in a deaerator in order to remove all dissolved air. Next the neutralization is carried out. The oil is heated in a countercurrent heat exchanger with the neutralized oil to 180°C and after a further heating to 250°C it is introduced to the highest section of the deodorizing column. There the carotenoids are destroyed and the fatty acids are distilled off at a pressure of 0.4–0.7 kPa using live steam. In the following section the oil is not heated. At a temperature of 240°C the remaining fatty acids and other volatile substances are removed. The oil flows through a heat exchanger to a cooler and leaves it at 50°C with the FFA content of 0.05% (Fig. 1.33).

A more thorough method of phospholipids removal from double-zero rapeseed oil enabled the application of physical refining for this raw material. The crude oil should not contain more than 70 mg/kg of phosphorus, while good results are obtained when this content is lower than 50 mg/kg. This is due to the fact that too high a phosphorus content causes an increased consumption of bleaching earth, which increases the costs of processing (Mag, 1983). The best method of phosphorus removal consists in an application of a mixture of citric and phosphoric acids. It is very important to thoroughly mix the acids with the oil, otherwise the phosphorus will be not entirely removed. Addition of water is important, since it facilitates the contact of polar phospholipids with the acids. After degumming, the oil is pumped to an apparatus, to which bleaching earth is also introduced. The suspension is subsequently pumped to a bleaching kettle, dried and bleached at a temperature exceeding 100°C. The course of the filtration requires that the oil is perfectly dry, otherwise the filtration rate decreases significantly. The bleaching earth should be removed from the oil, and the final phosphorus content should not exceed 5 mg/kg. The increase in the FFA content should not be higher than 0.2%. Physical refining makes the losses lower. The results of physical refining of 00 rapeseed oil achieved by the Canadian industry are the following: addition of phosphoric acid 0.07%, that of the bleaching earth — 1.5–2.0% at 110°C, deodorization at 260–270°C at a pressure of 670 Pa. The refined oil in the amount of 97.4%

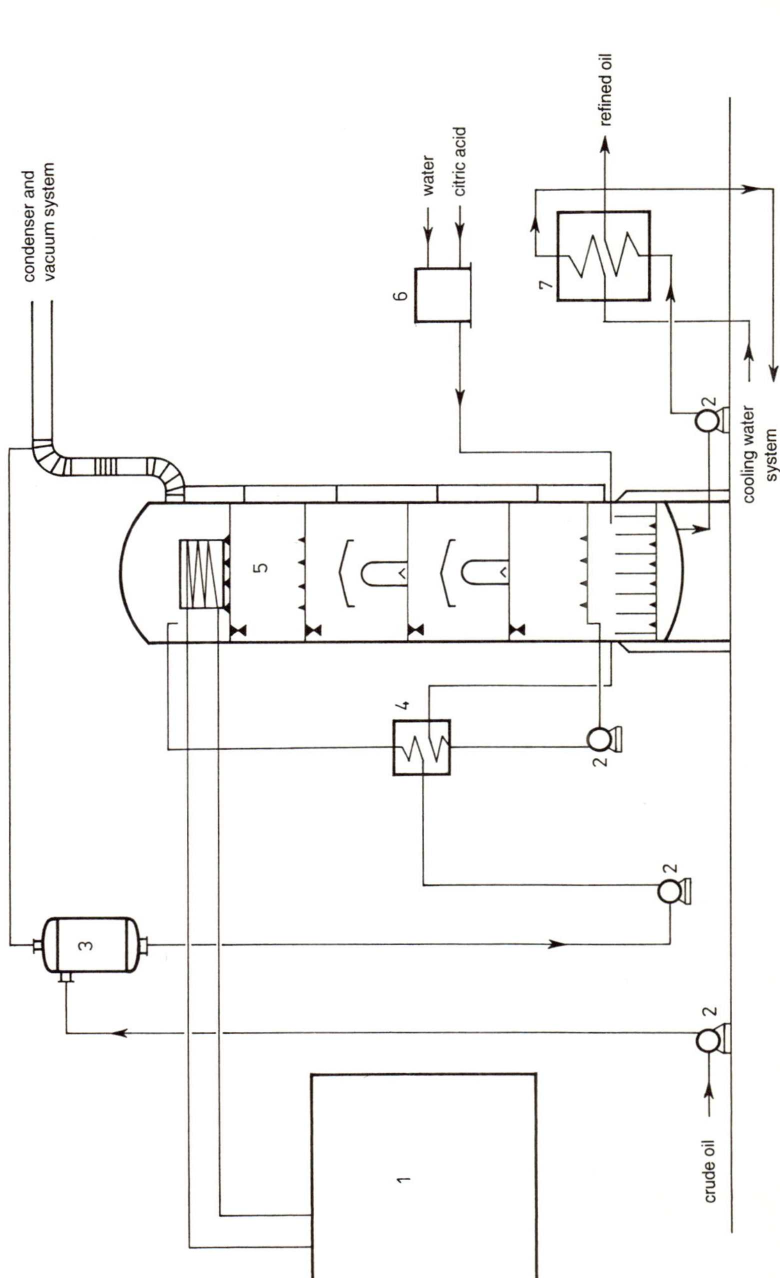

Fig. 1.33 Physical refining (Forster and Harper, 1983): 1—heat carrier container, 2—pump, 3—deaerator, 4—heat exchanger, 5—deodorizer, 6—citric acid feeder, 7—oil cooler.

of the initial quantity contained 0.028% FFA, had good colour and smell. The by-product was deodorization scum in an amount of 1.3% (Forster and Harper, 1983; Mag, 1983). The significantly higher FFA content in this by-product compared to the oil obtained from soapstock is worth noticing, since it testifies to the much smaller losses of neutral oil during physical refining. Apart from neutral oil, mono- and diglycerides also occur, being more volatile than triglycerides. The incomplete triglycerides are formed by a hydrolysis of the triglycerides during the refining or occur in the crude oil.

1.1.7.8 Neutralization in miscella

Neutralization in miscella eliminates washing with water, hence reduces the amount of wastes. However, this neutralization method is possible only in cases when the refinery is situated adjacent to the solvent extraction plant, and neutralization should be accomplished within 6 h from the extraction (Cavanagh, 1976).

1.1.7.9 Winterizing of oils

This operation aims at a removal of edible oil components causing its turbidity during a prolonged storage at a low temperature, e.g. in a refrigerator. The turbidity is usually due to crystallization of glycerides containing saturated fatty acid chains. In the case of sunflower oil the turbidity results also from the presence of waxes, which pass during processing from the hard hull to the oil. The percentage of hull varies from 28 to 35% of the kernel mass depending on the variety. Although the new varieties contain less hull, yet their wax content is much higher. Most of the waxes are removed during refining, but their residue should be removed more thoroughly during winterizing. The initial content of the waxes at the beginning of the winterizing process should be 0.17–1.33%.

Apart from waxes, also the glycerides. viz. stearodioleates and palmitodioleates, cause the turbidity of sunflower oil. Their total content is ca. 20%. Melting point of stearodioleates is 23.5°C, while that of palmitodioleates is 19°C. The winterizing temperature is usually 6°C and the process lasts for 6–18 h. As the system tends to a supersaturated state, an agent aiding the filtering is usually added at the beginning of cooling. This agent is usually silica added in an amount of 0.2–0.6%. After completing the filtration the cakes are heated to 77°C to melt the substance crystallized on the silica, so that it can be separated by blowing out. This method, however, is not suited for sunflower oil, which tends to polymerize already at 60°C. Another solution consists in extracting the cakes with a solvent, which is subsequently distilled off from the resulting miscella. A glyceride-wax fraction is thus obtained and the solvent is regenerated.

1.1.8 Hydrogenation of oils

Research on hydrogenation carried out in numerous centres not only elucidates the effect of the particular parameters on the properties of the hydrogenated fats produced, but leads also to a development of the equipment enabling the automation of the continuous process. Detailed information on the *trans*-isomer formation, changes in the IV and the thermal properties enables fats of various characteristics to be obtained. Basically continuous hydrogenation can be accomplished by using two kinds of catalyst. The first method is based on application of the catalyst in a suspended form (Fig. 1.34), while in the second a sprinkled stationary catalyst made of copper chromite is used (Fig. 1.35). The latter catalyst assures better selectivity (Mounts, 1983). Other advantages of the latter system are the lower reaction temperature and the formation of smaller amounts of conjugated dienes, the selectivity of linoleic acid being unchanged. It will be possible in future to obtain in such a way fats for special applications, or — using a traditional nickel catalyst — the components of margarine and other edible oils. Indepedently of the hydrogenation method, the problem of the spent catalyst always exists. When its activity is substantially reduced, it is removed from the apparatus after a very thorough removal of oil residues.

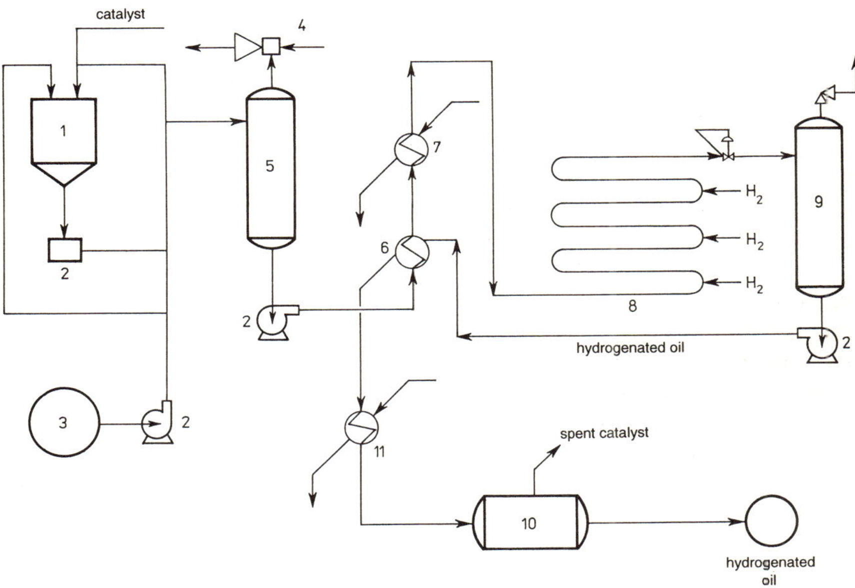

Fig. 1.34 Continuous hydrogenation with suspended catalyst (Mounts, 1983): 1—suspended catalyst container, 2—pump, 3—oil tank, 4—ejector, 5—deaerator, 6—oil heater, 7—oil superheater, 8—reactor, 9—degasifier, 10—catalyst filter, 11—cooler.

1.2 NON-EDIBLE FAT PROCESSING INDUSTRY

The non-edible fat processing industry is based to a large extent on raw materials which are by-products or wastes from other technologies. Non-edible fat processing is understood today to be the conversion of fats into fatty acids, glycerine and various types of derivatives of glycerides and fatty acids, often called oleochemicals. In these processes glycerides or

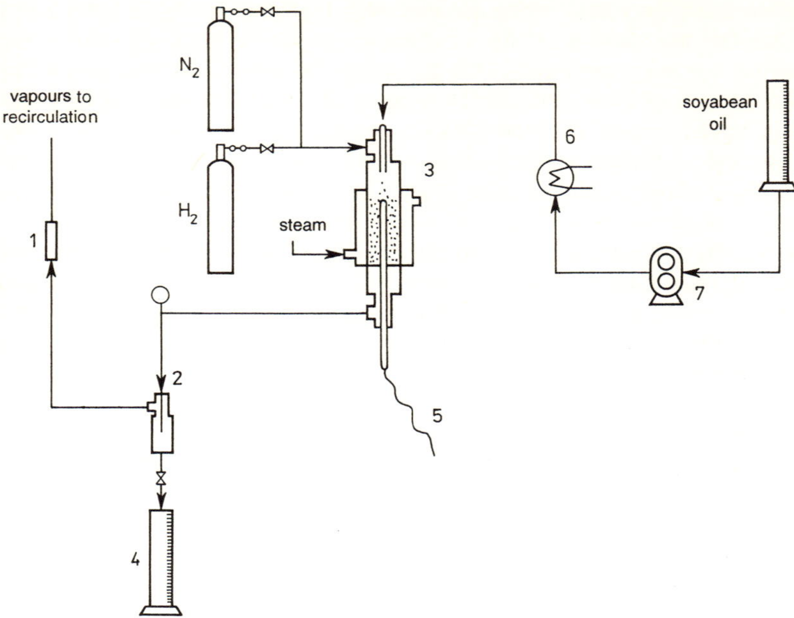

Fig. 1.35 Continuous hydrogenation with stationary catalyst (Mounts, 1983): 1—flow meter, 2—degasifier, 3—reactor with stationary catalyst, 4—hydrogenated product receiver, 5—thermocouple, 6—heat exchanger, 7—pump.

fatty acids constitute basic raw materials, although numerous auxiliary materials, both organic and inorganic, are also used.

Due to the deficit of food on a global scale, the raw materials for fat processing into non-edible products are increasingly often of poor quality — they are by-products or wastes from other industries (the manner in which they are formed is discussed in Section 1.3). Formerly the main fatty raw materials for obtaining non-edible products were full value beef tallow, as well as oils — coconut, palm, linseed, soyabean, and others, while recently the most important role is gradually being played by waste animal fats, vegetable post-refining acids, tall oil, marine animals' oils, utilization greases — by-products from the production of feed meals, sewer greases — from sewage treatment plants, degras — from tanning of leather, etc.

Non-edible fat processing develops in two directions, viz. as:

1. A rational method of utilization of fatty by-products and wastes formed in oil and other industries.
2. A set of technologies producing many new by-products and wastes which have to be rationally utilized or minimized.

The supply of raw materials and economic reasons promote the development of the first direction. The second is becoming increasingly significant in recent years due to the risk of damage to the environment. New technologies are faced with stringent requirements. The most favoured are those technologies that produce no wastes, with closed water systems and emitting no fumes, hence safe for the environment.

In the last fifty years the meaning of the term "non-edible fat processing" has significantly changed. Technological processes belonging to this group comprise not only splitting and distillation, saponification, separation into stearin and olein, hydrogenation of animal fats and bodying of drying oils, but also reduction of acids to alcohols, and nitriles to fatty amines, sulfonation and sulfation, esterification, oxyethylation, disproportionation, isomerization, polymerization and many other complicated syntheses of the so-called oleochemicals.

In all these operations by-products and wastes are formed. They are often of a non-fatty character. This chapter presents the more important processes of non-edible fat processing and the development noted which aim primarily at maximum utilization of the fatty matter, and thus a reduction of the amount of wastes.

1.2.1 Refining of non-edible fats

The raw materials for processing of non-edible fats are first of all by-products or wastes from other technologies. Their purity is usually low. The origin of these raw materials is discussed in Section 1.3, and their application — in Chapter 2. The impurities can be exogenous, like particulates, metal salts, sludge, pigments, water, traces of solvents, proteins, their decomposition products, etc.; or endogenous, like decomposition products of fatty matter — free fatty acids, peroxides, aldehydes, ketones, hydroxyacids and unsaponifiable matter. The fats are usually dark and have unpleasant smell. The majority of the impurities mentioned renders impossible the direct application of these raw materials for the production of soaps, lubricants, feed additives, oiling agents, etc., i.e. those products where the release of FFA from the glycerides is not necessary, or for which the presence of FFA is even undesirable.

The presence of impurities also hinders the processing of fats into DFA and glycerine, since it deteriorates the quality of the final products and significantly reduces their yield (Mazgajska *et al.*, 1978).

During fat splitting the protein substances and soaps form a sediment which sticks to the heating elements and hinders phase separation. Protein impurities can decrease the splitting degree by ca. 3% and can increase the glycerine losses by ca. 5%. Most of the organic impurities pass to sweet water and promote fermentation of glycerine during storage. They also cause an increase in the content of the organic residue in the water up to 15%, which increases the consumption of chemicals and the losses of glycerine in the sediment by ca. 100%. The quality of glycerine deteriorates.

During the distillation of fatty acids ca. 30% of the volatile products of protein decomposition pass to DFA and cause their rapid darkening. Non-volatile impurities increase the amount of the distillation residue up to 10%, thus decreasing the yield of the process.

Due to the above reasons the first and basic operation in non-edible fat processing is the refining proper for further processing.

There are a number of methods of non-edible fats refining. In the case of strongly contaminated waste fats a combination of several refining

Table 1.7 The technologies of non-edible fat refining (Szczepańska, 1983)

Type of refining	Principle	Main effect
Physical	settling centrifuging filtration heating to 240–280°C (degumming by mineralization) steam distillation liquid-liquid extraction	dehydration removal of solid impurities degumming
Chemical	action of concentrated or diluted acids (H_2SO_4, H_3PO_4) action of oxidizing agents (H_2O_2) degumming with chemicals (Na_3PO_4, Na_2HPO_4, NaOH)	removal of soaps and some oxidation products degumming and protein removal
Physico-chemical	adsorption on solid sorbents (asbestos, earths, active carbon) hydration (continuous in a column with NaCl solution or with steam in an ejector) hydrotropic action of solvents	degumming, removal of proteins or decolorization removal of volatile decomposition products
Mixed	hydrotropic degumming (hexane, isopropyl alcohol, H_3PO_4), adsorption on active carbon, chemical oxidation reduction of the oil/water interfacial tension using sodium benzene-, toluene- or xylenesulphonate, rinsing with NaOH solution, settling, drying	

methods is often applied. The general characteristics of these methods is presented in Table 1.7. Generally the refining methods are chosen accordingly to the composition of the raw material and the kind of the impurities, e.g. post-refining vegetable acids are not refined with concentrated H_2SO_4 in order to prevent the oxidation of the polyunsaturated acids.

1.2.1.1 Methods of animal fats refining for various applications

A non-edible fat is dissolved in a non-polar solvent, e.g. xylene, toluene, heptane (or their mixture) and is extracted at 15–70°C with an alcohol-water solvent (e.g. 5–23% methanol, ethanol or isopropanol). From 0.25 to 3 parts of the alcoholic mixture is used per one part of the non-polar solvent. After phase separation and distillation of the solvent, a refined fat is obtained from the non-polar phase, suitable for the production of lubricants, oils, detergents, or feed additives. After distilling the alcohol off a waste is obtained from the polar phase. This waste contains proteins, phospholipids, aldehydes, ketones and a part (2–5% of the total refined amount) of low molecular weight triglycerides (Łuczyn *et al.*, 1981). This manner of refining of animal fat and sewage grease allows the reduction of the phospholipid content of the refined fat by ca. 50%, almost entirely removes sludge impurities (to ca. 0.1%), reduces the FFA content by ca. 40% and the PV by ca. 50%. It also improves the colour and the smell of the fat.

If all the FFA are to be removed during the refining, the fat dissolved for example in petrol (1:4 vol.) is extracted with aqueous alcohol mixture containing ammonia (e.g. 30% H_2O, 66% ethanol and 3.3% ammonia). Multiple extraction (up to 4 times) is carried out at 45°C, the ratio of the polar solvent to the non-polar solvent being 1:1.5. The refined fat does not contain FFA, while the waste is enriched with them.

Utilization fat for feeding purposes is neutralized by superheated steam distillation, sometimes preceded by preliminary adsorption of the impurities on active carbon (Masłowska and Dorabialska, 1979). The distillation is carried out at 140–180°C (vapour temp. 125–140°C). The product is light in colour, of weak smell, the PV being reduced to 0 and the AV — by ca. 4 units. Similar results are obtained by chemical refining with oxidizing agents like hydrogen peroxide or potassium permanganate solution (Masłowska and Baranowski, 1979). The fat with 0.1% aqueous $KMnO_4$ solution acidified with H_2SO_4 is stirred for 0.5 h. Steam of a temp. of 100°C is passed subsequently through it for 10–40 min. After the separation of water and gum, the fat is washed with water and dried with superheated steam at 160°C (Baranowski and Busko-Baranowska, 1981).

Tallows of a colour exceeding 21 units in the iodine scale are bleached by the adsorption method before their use for the production of toilet soaps, since light colour of the raw material is necessary in order to obtain a proper appearance of the soap. Dark colour of tallows is mainly due to the presence of fat oxidation products, like hydroxyacids and ketoacids, and to contamination with protein-lipid complexes. Also metal ions from e.g. corrosion of the equipment can cause darkening of the fat, particularly in the presence of humidity and FFA. These impurities can be partly removed by an addition of 1–5% of bleaching earth. Two-stage purification at 90°C and 60°C is advantageous (Cegłowska and Turlewicz, 1979). Such a refining results in an improvement of the fat colour to the level of 5–12 iodine scale units. Oiled bleaching earth is a waste in this process.

Mink fat used as a lipophilic component of cosmetic products is purified by alkali refining aiming at the removal of FFA. The fat is neutralized with 9% NaOH solution in an amount 20% higher than stoichiometric. The smell is removed by deodorization at 205°C at a pressure of 10.6–13.3 kPa, lasting for 6 h (Volotovskaya *et al.*, 1973). Soapstock containing 22.8% FFA and 24.5% neutral oil is a waste from this process. It cannot be used for the production of soaps, but after mixing with spent bleaching earth it can be applied for example as a concrete plasticizer.

1.2.1.2 Methods of refining of fats for the production of fatty acids and glycerine

The simplest method of removal of impurities from a waste fat is a prolonged storage at an elevated temperature. This, however, requires a large storage area, and the final effect of partial separation of protein impurities, gums, phospholipids and soaps is achieved at the cost of a loss of glycerine — due to partial hydrolysis of fats during heating — and a deterioration of the colour of the raw material. The separated aqueous protein-lipid emulsions are wastes difficult to handle.

Particulates and gums are removed by filtration, using filter presses, drum or rotary filters, or arch sieves (Mazgajska *et al.*, 1978). A modern solution is the equipment for continuous decantation produced by the Westfalia company, removing 2–5% of water, as well as tiny solid impurities from the fat. At 3–5% content of water and solids an automatic liquid-solid phase separator can be used, while above 5% water content a liquid-liquid-solid separator must be applied (Dupjohann and Hemford, 1975). Application of continuous separators for rapid removal of water and gums from fats reduces glycerine losses, and the resultant waste contains less fat than that from the storage method. At 6000 r.p.m the optimal centrifuging time at 40–60°C is 20 min. Further elevation of temperature increases the fat content in water (Kamyšan and Derevjanko, 1972).

Waste fats can also be degummed in an electric field at a temp. of 100°C, or at 270°C under a reduced pressure passing an electric current of a frequency of 500–40000 Hz for 0.5–1 min. The fat is then cooled down and the coagulated sediment is separated. This method, however, causes changes in the fatty acids (Mazgajska *et al.*, 1978). To accelerate hydration of the impurities, the fat is mixed with superheated steam (9:1) at 135°C and passed through an ultrasonic field to a vacuum chamber (Ogilec and Ložešnik, 1981).

Three-stage refining of waste fats is recommended (Irodov, 1973). It consists in hydration with water or an NaCl solution, and subsequent action of diluted or concentrated sulphuric acid followed by bleaching with bleaching earth. Sulphuric acid decomposes soaps and reduces the amount of proteins and mineral salts in the fat. Concentrated sulphuric acid under properly chosen conditions causes:

— mineralization of protein-type organic substances, resulting in a substantial or even total removal of nitrogen compounds,
— decomposition of metallic soaps and transfer of metal ions to the aqueous phase, which reduces the metal content of the fat (mainly that of iron and calcium),
— decomposition of inorganic salts, e.g. phosphates,
— oxidation of peroxides to hydroxyacids, and aldehydes to acids,
— improvement of the smell, reduction of the ash content, decrease of the susceptibility to oxidation — by a removal of oxidation catalysts.

An additional effect of hydration of gums occurs during the removal by washing of the sulphuric acid (Szczepańska *et al.*, 1972).

The main disadvantage of application of sulphuric acid is the formation of esters — with free hydroxyl groups of partial glycerides, and of oxysulfoacids — by addition to unsaturated bonds of fatty acids. Due to their surface-active properties, these compounds can cause emulsification of the fat with water, since their hydrolysis with evolution of sulphuric acid is very slow (Volotovskaya *et al.*, 1979). Sulphuric acid can cause the formation of high-molecular fat oxidation products, which accumulate in the distillation residue. Washing the sulphuric acid out requires large amounts of water, hence much waste is formed in this method (Gavrilko *et al.*, 1972). Despite these disadvantages, purification with sulphuric acid is most often used for refining very contaminated fats. In modern practice the contact time of the fat with H_2SO_4 is shortened due to vigorous mixing of the two phases accomplished by means of e.g. a static mixer. The complete mixing of the fat with the acid is achieved in 80 s (Mazgajska *et al.*, 1979). Multistage refining of waste animal fats including the sulphuric acid refining stage is presented in Fig. 1.36. The process can be carried out in a continuous manner using steam jets for hydration, centrifuges and Fund type filters for degumming and adsorbent removal, and a static mixer for H_2SO_4 refining. Not all the impurities are removed in this process — the remainder is distributed among the particular intermediates.

Instead of sulphuric acid, phosphoric acid can also be used for the refining of waste fats. The fat is mixed at 75–80°C with 0.1–0.2% H_3PO_4 for 20–30 min. Then the mixture is left to settle the gums, the acid is washed out with water at 85–95°C (the amount of water is equal to 10% of the initial fat mass), and finally the fat is refined by adsorption (Volotovskaya and Grin, 1982). The effects of such a refining process are listed in Table 1.8. Slightly better effects are achieved when the H_3PO_4 refining is followed by extraction with aqueous solution of metal (sodium or potassium) salts and monobasic acids, and subsequent separation of the aqueous phase. The extraction can be carried out using for example a 3% potassium oleate solution at 65°C, the pH of the aqueous phase being 6–9. Waters from washing the oils after alkali neutralization can be used for the extraction (Volotovskaya *et al.*, 1982).

Waste fat refining methods without acid refining are also used. A properly selected surfactant is added to the refined fat. The method consists in increasing water solubility of fat impurities by reducing the surface tension of water and increasing the hydrophilic character of the impurities. Dissolution of small amounts of surfactans in water reduces its surface tension two to three times. Aqueous surfactant solutions are adsorbed at the water-fat and water-impurity interfaces, thus reducing the interfacial tension. A proper selection of the surfactant enables reducing the interfacial tension at the water-impurity interface more than at the water-fat interface. This increases the hydrophilic character of the impurities, their wettability, and the degree of dispersion in the aqueous phase. The resultant emulsion should be unstable and should be easily breakable into water and fat. Apart from the kind and concentration of the surfactant, also the temperature, stirring intensity and the ratio of the phases influence this process. In the continuous method of fat refining by means of surfactants the raw fat and the aqueous surfactant (e.g. sodium alkylsulphonate) solution in a proportion of 25–50% of the aqueous phase to the fat are heated to 90°C and mixed in a continuous mixer for ca. 10 min. Then the phases are separated in a separator. The fatty phase is washed with water at

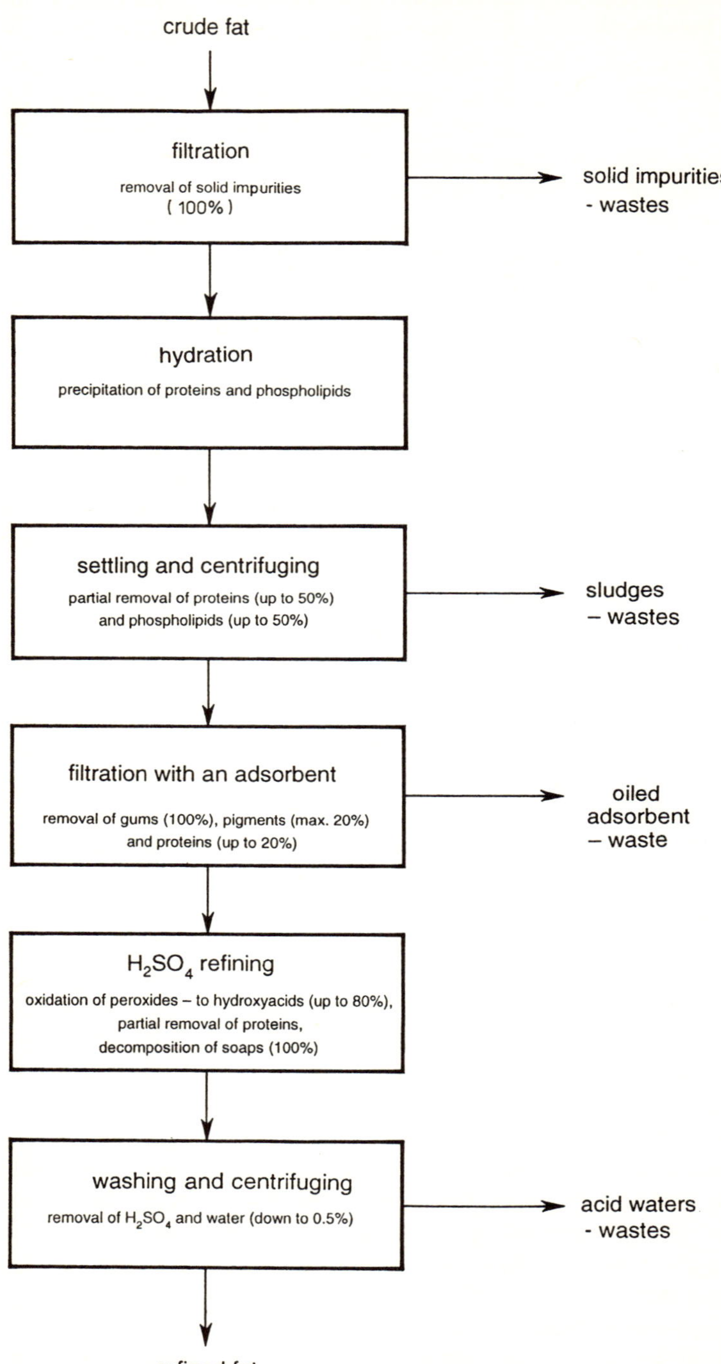

Fig. 1.36 Refining of waste fats for DFA production (Mazgajska *et al.*, 1978).

Table 1.8 Characteristics of fat before and after refining with phosphoric acid and bleaching earth (Volotovskaya and Grin, 1982)

Components / Properties	Content (%) in fat	
	crude	refined
Ash	0.037	traces–0.012
Phosphorus	traces	—
Nitrogen	0.08	0.03–0.05
Ether insolubles	1.44	0.38–0.75
AV	8.5	8.8–9.7

90°C(Gavrilko*et al.*, 1976). Fat refined under optimal conditions contains 0.08% proteins, 0.05% phospholipids and 0.03% soaps. The waste contains 4–5% non-fatty impurities and 1.5–2% surfactants. Its amount is equal to 25% of the fat amount. If the chosen surfactant is biodegradable, the waste is passed to a waste treatment plant. Water from washing (or the condensate) contains 0.1–0.2% non-fatty impurities and 0.2–0.3% surfactant. It is recycled to the preparation of the surfactant solution.

1.2.2 Hydrolysis of fats

Hydrolytic splitting of fats is commonly used to isolate fatty acids from non-edible fats. The main products of this process are raw fatty acids (SFA), while glycerine is a by-product. Fat splitting aims primarily at yielding a maximum splitting degree expressed as the AV/SV ratio. When the splitting degree reaches 100%, the further SFA processing proceeds with negligible losses of fatty matter in the distillation residue. The yield of the by-product — glycerine — also increases.

The amount of SFA that can be obtained from hydrolysis of any fat is equal to:

$$\%\mathrm{SFA} = \frac{3M\ [100 - (B+N+K)]}{3\,(M-1)+41} + (N+K)$$

where M — fatty acid molecular weight, B — percentage of water in fat, N — percentage of unsaponifiable matter in fat, K — percentage of FFA in fat.

During high-pressure fat splitting ca. 0.03% of fatty acids is lost due to decomposition, 0.002% is lost with the expander fumes and 0.05% with sweet waters (Irodov, 1970). In practice the splitting degree in high-pressure hydrolysis reaches 99% in the continuous process and ca. 94–97% in the batch process, the results depending both on the kind of fat and the manner of refining prior to splitting, and on the process and equipment used.

Fat hydrolysis is a reversible reaction that proceeds gradually, since the fatty acid radicals are liberated one by one from the triglycerides and the products of incomplete hydrolysis, viz. mono- and diglycerides, are formed.

Hydrolysis is a first order reaction proceeding in the fatty phase, mainly between the fat and water dissolved in it. Interfacial phenomena are believed to be of minor importance (Anon, 1973). The initial reaction rate during hydrolysis of fats of low FFA content is low. This induction period results from the poor solubility of water in neutral fats. Hydrolysis in a heterogeneous systems prevails in this period. When the FFA content reaches 15–20%, the reaction rate increases, since water solubility in fatty acids is higher than in neutral fat. Therefore the induction period in hydrolysis is influenced by:

— hydrolysis temperature, the increase of which improves water solubility in fat and fatty acids (Tables 1.9 and 1.10),
— FFA content in the split fat,
— incomplete glycerides content in the split fat (due to a reduction of the interfacial surface tension) (Table 1.11).

Table 1.9 The effect of temperature on the time of fat hydrolysis necessary for reaching equilibrium (Anon, 1973)

Temperature (°C)	Time necessary for reaching the equilibrium (min)		
	beef tallow	coconut oil	peanut oil
225	156	158	156
240	82	84	85
260	47	46	53
280	34	33	33

Table 1.10 Solubility of water in fatty acids (Molčanov and Evdokimov, 1971)

Temperature	Amount of dissolved water (% w/w)*	
(°C)	FA of hydrogenated sunflower oil	animal FA**
77	0.9	1.1
102	1.4	1.8
163	3.7	3.7
222	8.3	9.2
248	11.8	—
268	15.9	16.8
283	—	21.0
294	—	26.0

* Solubility of water in triglycerides at 260°C is 6% (Basu, 1976).
** 94% splitting degree.

The reaction equilibrium is influenced mainly by the concentration of glycerine in the fatty phase (Basu, 1976). Glycerine is practically insoluble in dry fat. However, when the fat contains dissolved water, the solubility of glycerine distinctly increases. After mixing with a 15% glycerine solution

Table 1.11 The effect of FFA and partial glycerides on fat hydrolysis process ($t = 225°C$, $p = 2.5$ MPa, time 3 h), (Butenev *et al.*, 1980)

FFA (%)	Partial glyceride content (%)	Splitting degree (%)
5.7	—	80.2
10	—	84.4
16	—	86.8
—	7.2	85.0
—	10.1	90.1
—	15.0	90.5

Table 1.12 Fat hydrolysis and esterification rate constants (Basu, 1976)

Reaction	Reaction rate constant at 260°C	
	K	K'
Triglyceride + water $\underset{K'_{I}}{\overset{K_{I}}{\rightleftharpoons}}$ diglyceride + fatty acid	5×10^{-4}	2×10^{-4}
Diglyceride + water $\underset{K'_{II}}{\overset{K_{II}}{\rightleftharpoons}}$ monoglyceride + fatty acid	5×10^{-4}	7×10^{-4}
Monoglyceride + water $\underset{K'_{III}}{\overset{K_{III}}{\rightleftharpoons}}$ glycerine + fatty acid	6×10^{-4}	7.5×10^{-4}

from 1 to 5% of glycerine dissolves in fat containing 10% water (Sikorski and Staniewski, 1976). Hydrolysis and esterification rate constants are listed in Table 1.12.

Hydrolysis degree in equilibrium (H) is determined by the so-called Lascaray formula (Anon, 1973):

$$H = 100 - 8 \times 10^{-1} G$$

where G — glycerine concentration in sweet water. The dependence is limited to G values lower than 20%.

On the basis of the kinetics of fat hydrolysis it has been possible to determine the most advantageous technology and equipment for this process, starting from low temperature catalytic processes in which the catalyst played the role of an emulsifier, to high temperature continuous countercurrent processes, carried out even at 260°C. It enabled a significant shortening of the induction period and a displacement of the equilibrium to a splitting degree of 99%. Colgate-Emery or Lurgi systems are examples of such systems. The basic element of such installations is an 18–26 m

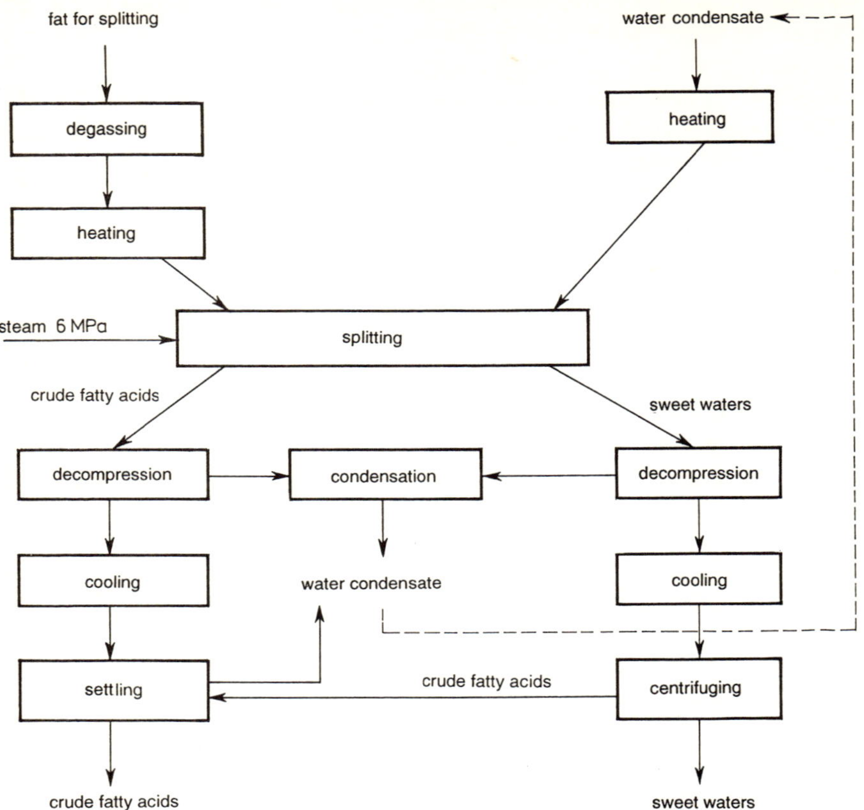

Fig. 1.37 Outline of high-pressure continuous fat splitting process (Mazgajska and Smenda, 1978).

reaction tower, fed by high-pressure pumps supplying countercurrently fat and water. A schematic diagram of such an installation is presented in Fig. 1.37. The reactor operates at 250–260°C under a pressure of 5.4–6.0 MPa. Depending on the kind of the raw material, it remains in the reactor from 1.6 to 3 h (Mazgajska and Smenda, 1978). The raw material in this method must be very pure in order to avoid the formation of emulsions. The concentration of sweet waters reaches 20% (Szczepańska *et al.*, 1982), and the splitting degree is very high. As the hydrolysis temperature is high, certain side reactions resulting in an increased amount of the SFA distillation residue can take place in this method (Efimov *et al.*, 1980).

Other equipment for the splitting process consists of a reactor cascade operating countercurrently, e.g. Buss equipment (Efimov *et al.*, 1980) or Gianazzi (Grecki, 1961).

The fat splitting process is also carried out in batch systems, particularly in the case of non-uniform raw materials or small capacity systems. In the majority of modern batch processes the splitting is performed under a pressure of ca. 3 MPa, without any catalyst, in two stages. In the first stage the hydrolysis is carried out in the presence of a certain amount of circulating water and a splitting degree of up to 90% is reached. Next the sweet water is separated and a new portion of water is added to the SFA. The maximal splitting degree in this method is 96%, and the concentration of the sweet waters is equal to 10–12% depending on the kind of the raw material and the amount of water. Post-refining acids split with this method under a pressure of 1.8–2 MPa at 210°C for three hours yield after the first stage a splitting degree of 85–90% and the concentration of sweet waters equal to ca. 5%, while after the second stage — 94–96% and only 1.8–2.0% , respectively (Rojzman *et al.*, 1971). An increased amount of water added at the particular stages increases the splitting degree, but at the same time causes dilution of the sweet waters. High pressure hydrolysis of fats requires expensive equipment. Therefore proposals of new technologies have recently been put forward, in which a hydrolysis temperature of 200°C is achieved without the use of high pressure by the application of superheated steam. The process takes place in a tube reactor in the presence of zinc oxide as a catalyst. The evolved glycerine is removed from the reaction medium in gaseous form. A splitting degree of 16% is achieved after the first run of the tallow through the reactor at 200–280°C within 10–30 s, while after the second run it is 70% (Sonntag, 1979 a, b, 1984). Some technical raw materials, e.g. castor oil, require special hydrolysis conditions, since an excessively high temperature can cause dehydroxylation of the ricinoleic acid. In such cases the upper temperature limit is 210°C, and the optimal splitting conditions for a two-stage splitting are: 2 MPa pressure, 130% of water and the time of 4 h (Witwicka *et al.*, 1979).

The process of high-pressure non-catalytic fat splitting is almost waste free. Fat soluble impurities present in the raw materials pass to the SFA, and the water soluble to the sweet waters. These impurities are removed during subsequent refining stages (FA distillation and purification of sweet waters). The fumes from the splitting installation can constitute a hazard for the environment (Irodov *et al.*, 1971).

In high pressure splitting decomposition of fats and the accompanying substances (e.g. proteins) occurs due to the action of heat and steam. Volatile decomposition products which do not condense with steam are waste fumes of a disagreable odour. Waste gases from the expander in an amount of 1.5–3% of the SFA can be directed to tube condensers of large surface area, where the majority of the gases is condensed. Water condensate containing some of the condensed fumes can be recycled to splitting.

A future solution for industrial fat hydrolysis is enzymatic hydrolysis (Sonntag, 1984). The process takes place at 40–60°C under the conditions of low SFA corrosivity, which reduces the costs of the equipment and energy, and eliminates the problem of thermal destruction of the raw material. Lipase is used as the enzyme. The source of lipase is important, since it should not exhibit any selectivity with respect to the position in the glyceride and to the kind of the acid moiety (Posorske, 1984). Highly active lipase obtained e.g. from *Candida rugosa* exhibits such properties (Linfield *et al.*, 1984). Lipase obtained by powdering the degreased ricinus grain can also be utilized. Such a preparation is added in an amount of 10% of the fat mass. The hydrolysis is carried out at pH 4.8 in the presence of acetic acid (60% of the fat mass) at 40°C for 5 h on stirring. A splitting degree of 87–95% can be achieved depending on the kind of raw material (Table 1.13). The process can be carried out in a continuous manner, using a tube reactor. After the process the hydrolysis product is heated to 80°C, and the SFA, the enzyme and the sweet water are centrifuged.

Table 1.13 Results of enzymatic fat splitting by means of ricinus seed lipase (in an amount of 10% relative to the mass of the fat) in a CH_3COOH medium (60%) at pH 4.8 within 5 h, at 40°C (Tresko *et al.*, 1977)

Raw material fat (grease)	AV	Splitting degree
Bone	56	95.9 ± 0.6
Utilization	46.2	88.2 ± 0.5
Pork, cutaneous	6.2	87.5 ± 0.6
Lard	0.2	94.7 ± 0.7
Crude cottonseed	9	88.8 ± 0.05
Refined sunflower	0.2	93.9 ± 0.8

1.2.3 Fatty acids distillation

Distillation under a reduced pressure is used for refining raw fatty acids. Special distillation conditions must be observed, since fatty acids are very susceptible to oxidation and thermally induced changes. SFA contain ca. 0.02% dissolved air, an average of 1–3% water, 0.01–0.1% highly volatile components, and impurities: unsaponifiable matter, non-volatile unsplit glycerides, as well as fatty acid and glyceride polymerization and polycondensation products (Stage, 1979).

Low-boiling substances originate mainly from the decomposition of fatty acids. They are low-molocular weight hydrocarbons, aldehydes and ketones. The amount of the intact glycerides in SFA depends on the course of splitting prior to distillation. Polymerization of glycerides and fatty acids proceeds under the influence of temperature changes. It is often catalyzed by the impurities. Vegetable acids of high iodine value are particularly susceptible to such changes.

Distillation of fatty acids aims at efficient separation of them from all the other types of chemical compounds. The distillation should not yield new impurities. Therefore distillation conditions should be chosen so that:

— as much as possible of the volatile SFA impurities are separated in the form of forerunnings,
— the unsplit fat, UM, polymers and oxyacids are separated in the distillation residue,

—oxidation, decarboxylation and polymerization are avoided during the distillation.

The usual designs of the equipment aim at:

—a removal of the dissolved air, steam and a part of the volatile impurities in the so-called preliminary degasser,
—an increase in volatility of the fatty acid vapours achieved by reducing the pressure and by injection of steam,
—minimization of the residence time of fatty acids at high temperature, as well as cooling them down before leaving the low pressure zone,
—application of an inert gas during operations with highly unsaturated fatty acids,
—selection of proper materials resistant to fatty acids — not only for the sake of the equipment, but first of all for the sake of the fatty acids, since e.g. trace amounts of iron in FFA can cause considerable changes due to decomposition (oxidation).

Degassing should be carried out at a relatively low temperature under a pressure of 100–200 hPa. Under such a pressure the condensation temperature of the fumes is 51–66°C and membrane condensation can occur, which prevents environmental pollution (Stage, 1975b). If this stage proceeds improperly, the DFA are darker and have a disagreable odour. Figure 1.38 presents the vapour pressure curves for fatty acids and volatile decomposition products vs. the temperature. The distillation is carried out under a pressure of 3–10 hPa achieved by using multistage steam ejectors with membraneless condensation. The steam is usually injected directly to the distillation retort in order to decrease the boiling point (Fig. 1.39).

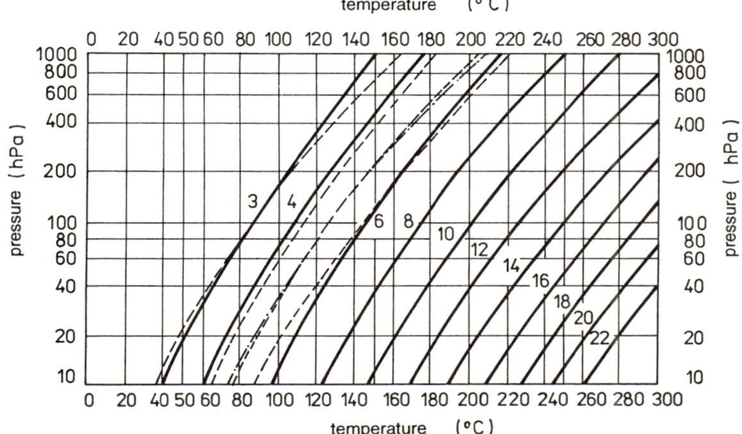

Fig. 1.38 Pressure curves of vapours of low boiling impurities and fatty acids depending on temperature (Stage, 1975b). Solid lines — low-boiling fatty acids (3–22°C), dashed lines — low-boiling methyl ketones (8–10°C), dot-and-dash line — low-boiling nonyl aldehyde.

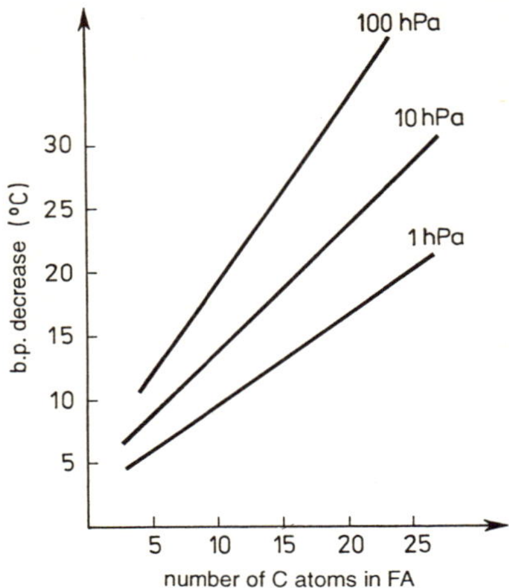

Fig. 1.39 Decrease of fatty acid boiling temperature resulting from an addition of 10% w/w steam directly to the distillation retort, depending on the number of carbon atoms in the acid and on pressure (Stage, 1973).

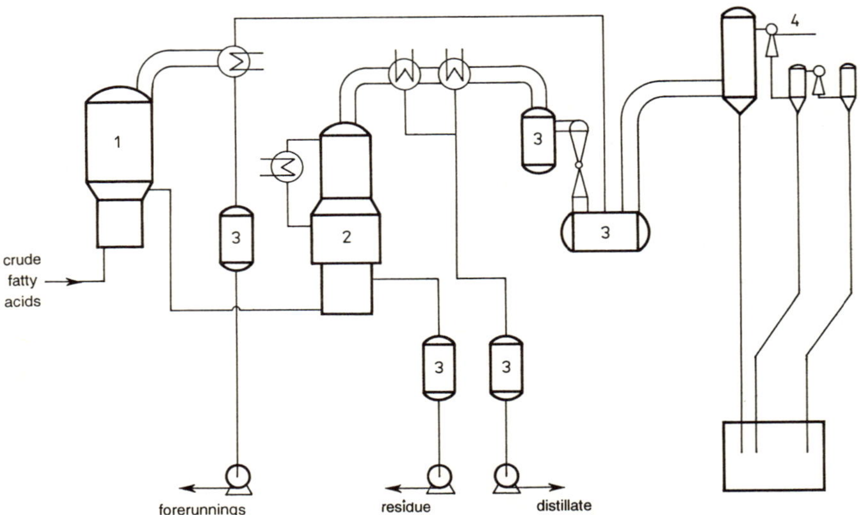

Fig. 1.40 Lurgi system for continuous distillation of fatty acids: 1—forerunnings distiller, 2—FA still, 3—tank, 4—vacuum system.

The upper temperature limit in distillation of saturated fatty acids is 265°C for 10 min. In the case of unsaturated acids the temperature should not exceed 240°C. Falling film type evaporators, in which the temperature of the heating medium is kept not more than 20°C above the maximum temperature in the evaporator, are recommended in order to assure a short residence time at the high temperature. Figure 1.40 presents an example of the equipment for the distillation of fatty acids (Lurgi company).

Distillation of fatty acids is not carried out to a complete removal of FA from the residue. An AV of the residue equal to 50 is a safe limit. This is a so-called residue I, which is a by-product of the distillation. If a two-stage distillation is carried out, so-called forerunnings are obtained from the retort I. They also constitute a by-product of this process. For example, during distillation of SFA obtained from utilization grease the amount of the residue I can be ca. 16% with respect to SFA, while that of residue II is ca. 4% (Szczepańska *et al.*, 1971). The yield of distillation of post-refining vegetable acids and highly unsaturated acids, e.g. from linseed oil, as well as dehydrated acids — from ricinus oil, can be much lower, down to 50% in cases of considerable polymerization of the raw material (Leščenko and Levit, 1973).

Certain reagents preventing the changes, e.g. natural antioxidants, can be used to reduce the intensity of side processes (Levit *et al.*, 1971; Table 1.14).

Table 1.14 The effect of anti-oxidant on the results of fatty acid distillation (Levit *et al.*, 1971)

Raw material SFA	Inhibitor addition*	DFA yield (%)	Residue (%)	DFA IV
Dehydrated castor	—	50.4	37	143
oil	0.5%	60.7	35.7	148
Linseed oil	—	75	20	172
	0.5%	77.1	16	168

* Nitrogen gossypol derivatives were used as the inhibitor.

Losses of fatty acids and products of their decomposition during the distillation are the most dangerous for the environment.

Large amounts of cooling water (60–80 m^3/t DFA) are consumed during the FA distillation. The water is passed to a barometric condenser. Part of the emulsified fatty matter can be separated in a fat trap; the rest passes to wastes. In such cases the wastes contain volatile distillation products — formic and acetic acids, acrolein, aldehydes and ketones, as well as large amounts of entrained or emulsified high-molecular weight acids (Table 1.15). The content of these substances in wastes depends on temperature changes of the cooling water, as well as the amount and pressure of steam injected to the distillation retorts. The total amount of the live steam should be equal to 6.5–7.5% with respect to the DFA (Irodov, 1976). The temperature of water in the barometric condenser should be maintained at 30–32°C. Losses of FA in the wastes can be reduced to 1–1.5 kg/t DFA by observing the above conditions.

Stage (1975) recommends a closed water system utilizing plate heat exchangers. In this system the amount of water contacting the fatty acids at a throughput of 5–10 t DFA/h does not exceed 1 m^3/h. Traces of fatty

Table 1.15 Composition of fatty acids in wastes from FA distillation (Irodov *et al.*, 1976)

Acid	Content (%) in wastes from distillation of	
	bone fat FA	vegetable post-refining acids
C_6	traces	—
C_7	1.1	traces
C_8	3.2	0.8
C_9	1.0	1.0
C_{10}	0.8	1.8
C_{11}	traces	1.9
C_{12}	1.5	2.8
C_{13}	traces	0.5
C_{14}	2.3	2.4
C_{15}	traces	0.3
C_{16}	25.4	19.3
$C_{16:1}$	5.2	3.9
C_{17}	2.4	6.7
$C_{17:1}$	traces	6.7
C_{18}	15.7	6.4
$C_{18:1}$	35.6	38.9
$C_{18:2}$	5.6	7.8
C_{20}	—	2.2

acids, aldehydes and ketones at the level lower than 10 mg/kg are found in the wastes.

1.2.4 Fractionation of fatty acids

The increase in industrial production of oleochemicals is accompanied by an increasing demand for fatty acids. The annual increase in production of fatty acids has been estimated in the seventies to be equal to 3% (Leonard and Kopald, 1984). Growing attention is paid at the same time to proper utilization of the specific properties of particular groups of fatty acids. Industrial methods of FA fractionation serve for this purpose. Fatty acids differing in their melting points are separated into liquid and solid fractions by crystallization from solvents or from aqueous disperse systems (the so-called hydrophilization method). Fractional distillation of fatty acids or their methyl esters is usually applied for fractionation of fatty acids with respect to their molecular weight (Table 1.16).

The most frequently produced fatty acids are: oleic acid ($C_{18:1}$), stearic acid (C_{18} or a mixture with C_{16}), palmitic acid (C_{16}), lauric acid (C_{12}), myristic acid (C_{14}), as well as erucic acid ($C_{22:1}$) or behenic acid (C_{22}), and also unsaturated C_{18} acids.

Owing to large differences in the properties of FA, valuable chemicals of various applications can be produced by fractionation and further processing of these acids.

Table 1.16 Boiling point (°C) of FA and their methyl esters (Cechnicki and Kośmicki, 1970)

Number of carbon atoms in FA chain	Boiling point at a pressure of (hPa)					
	1.3	5.2	13	1.3	5.2	13
	fatty acids			methyl esters		
10	110.3	132.7	148.8	66	89.3	106.6
12	130.2	154.1	172.0	92.3	115.9	134.0
14	149.2	173.9	192.0	114.8	140.0	160.8
16	167.4	192.2	210.7	137.0	162.9	183.8
18	183.6	209.2	228.7	158.0	185.3	206.0
20	200.0	227.3	248.0			
22	214.0	242.0	263.0			

Depending on the fractionation method applied, the by-products, the wastes and the degree of risk to the natural environment are different.

1.2.4.1 Fractional distillation

This method is usually applied for the fractionation of coconut acids, palm kernel acids, high-erucic rapeseed acids or their methyl esters, fish oil or hydrogenated fish oil fatty acids, and more seldom for animal fat acids. If the FA fractions can be used in later stages of processing in the form of methyl esters, it is recommended that the methyl esters obtained by transesterification of fats are fractionated instead of the acids. Methyl esters have lower boiling points than the respective FA and the differences in the b.p. of the homologues are greater; moreover, the esters are more thermally stable than the respective FA, do not form anhydrides at high temperatures and are not aggresive towards the materials the equipment is made of (Sonntag, 1984). Two kinds of FA fractionation are used: so-called partial fractionation yielding a chief product and by-products, and complete fractionation aiming at obtaining several products. In the latter case only the distillation residue or the so-called intermediate fraction are by-products of the process (Berger and McPerson, 1979).

Distillatory fractionation creates hazards similar to those in the previously described distillation process:

— thermal decomposition of the raw material resulting in an increased amount of volatile compounds and the possibility of environment pollution,
— thermal degradation of the carboxylic group and the unsaturated bonds (formation of hydroxyacids, polymers), causing an increase in the amount of wastes.

The hazard is lower when saturated (e.g. hydrogenated) acids are fractionated, since they are less susceptible to oxidation. Fractionation of unsaturated acids requires additional precautions in the design of the equipment.

Fractional distillation of fatty acids was traditionally carried out in plate or packed columns with injection of live steam, aiming at a reduction of the degree of thermal degradation of the FA (Stage, 1984a; Table 1.17).

Table 1.17 The effect of live steam on the reduction of temperature in the head and inside the distillation column during the separation of C_{16} and C_{17} FA at a pressure in the head equal to 5 hPa and a 10 hPa pressure drop in the column (Stage, 1984a)

Live steam (% w/w)	Reduction of temperature (°C)	
	column head C_{16}	column interior C_{17}
1	3	7
2	5.5	7
5	12	9.5
10	19.5	29.5
20	30	44.5

In such a process the condensation of volatile fatty acids can be incomplete and the uncontrolled losses can be increased, which is dangerous to the environment (Table 1.18).

Table 1.18 Fatty acid losses during distillation of C_{12}–C_{18} FA at 4 hPa pressure in the column head, depending on the application of live steam or air and the temperature of the condenser (Stage, 1973)

Number of carbon atoms in FA	Temperature of the condenser (°C)	FA losses (kg/h) due to application of		
		live steam 100 kg/h	live steam 50 kg/h	air 5 kg/h
12	100	30.5	15.3	1.91
14	100	8.3	4.2	0.52
16	100	2.1	1.0	0.134
18	100	0.4	0.2	0.025
12	80	5.4	2.7	0.336
14	80	1.6	0.8	0.096
16	80	0.36	0.18	0.022
18	80	0.08	0.04	0.05
12	60	0.84	0.42	0.052
14	60	0.19	0.096	0.012
16	60	0.042	0.021	0.0027
18	60	0.012	0.006	0.0007

In modern procedures the injection of live steam is not used. High efficiency of the process is achieved by a proper design of the columns, assuring a small pressure drop. Moreover, evaporators assuring short residence time of FA at high temperature and vapour condensation systems of a suitable design are used (Stage, 1984b). The principal conditions of efficient fractional distillation are:

— thorough degassing and dehydration of FA,
— moderate thermal conditions during evaporation,
— proper column packing assuring a high separation efficiency at a low pressure drop,
— anti-corrosive protection of the equipment,
— complete protection against air or wastes pollution.

Degassing and dehydration are performed at 150–180°C at a pressure of ca. 100 hPa (Stage and Bose, 1974).

Vaporization of FA is carried out in rotating thin film-type evaporators, the limitations being the same as in the case of simple distillation. The dependence of the maximum allowable boiling temperatures on the FA iodine value is presented in Table 1.19. These temperature limits determine the pressure in the evaporator for fractionation of FA.

Table 1.19 Maximum permissible boiling point depending on the FA iodine value and time (Stage, 1984a)

IV range	Maximum permissible boiling point (°C) within		
	40 min	20 min	10 min
< 60	245	250	260
60–140	235	240	245
> 140	225	230	235

The design of the distillation columns should assure a pressure drop securing safe operating conditions taking into account the pressure in the column head. Depending on the kind of the packing, the pressure drop can be equal to 0.1–2 hPa per one TP. The number of the TP depends on the fractionation degree required and usually does not exceed 20–30 (Berger and McPerson, 1979).

The material for the parts of the equipment operating at high temperatures is selected depending on the kind of raw material being fractionated. If FA with a number of carbon atoms equal or higher than twelve are fractionated, the molybdenum content of the stainless steel should be at least 2.2% ; when lower molecular weight acids are present, the molybdenum content should be 3–4% (Stage, 1984b). The corrosive action of fatty acids is stronger in the presence of traces of air (Stage, 1978).

When fractional distillation is carried out without the use of live steam, only very small amounts of FA reach the vacuum units. In an installation of a throughput of 6 t/h the amounts of steam and air reaching the vacuum unit are 8 and 12 kg, respectively. When the outlet temperature is 25°C, the fumes contain ca. 0.02 kg/h of fatty acids. Assuming that the water consumption in the vacuum unit is 4.2 m^3/h and that all the acids are dissolved in the cooling water, the FA concentration in water from the barometric condenser is not dangerous for the environment (Stage and Bose, 1974).

Fractionation of coconut FA can yield individual acids of ca. 99% purity. The number of TP for this purpose should be ca. 20. The number of columns necessary depends on the throughput. Stage (1979) recommends a two-column system at a throughput of 8–24 thousand t/year, 3 to 4 columns at a throughput higher than 24 thousand t/year, and one at a throughput up to 2 thousand t/year. In order to completely separate the FA, the mass flow through a single column should be repeated five times, while through two columns — twice. Fractional distillation conditions are listed in Table 1.20.

Table 1.20 Temperature and pressure during fractional distillation of coconut oil acids (99% purity fraction) (Stage, 1978)

FA distillate number of carbon atoms	Conditions in the head of the column (plate and packed)		Conditions at the bottom of the column			
			plate column		packed column	
	pressure (hPa)	temp. (°C)	pressure (hPa)	temp. (°C)	pressure (hPa)	temp. (°C)
6–10	13.3	130	77	224	26.8	198
12	13.3	172	77	246	26.8	218
14	13.3	193	69	262	26.8	237
16	6.7	197	62	265	20.0	236
8	53.4	153	117	199	66.3	184

Fractionation of high-erucic rapeseed acids is much more difficult due to high boiling point and high IV of these acids. When obtaining erucic acid of 95% purity, free from decomposition products, the number of theoretical plates in a column is limited to a large extent by the allowable pressure drop. When the pressure in the column head is 2 hPa, the number of the TP can be 20 at the most in order to maintain preservative conditions of FA distillation. In the ACV Company 4-stage installation for fractional distillation the FA of high-erucic rapeseed can be separated into chief and by-products (see Table 1.21).

Pure palmitic acid is obtained from cotton-seed acid fractionation or hydrogenated acid fractionation in a two-column installation: the first bubble-cap column of 24 TP serves for the separation of the forerunnings, while in the second packed column pure palmitic acid is obtained (Kločko and Vetkazova, 1976). The conditions and the results of fractionation are listed in Table 1.22.

Table 1.21 Results of fractionation of high erucic rapeseed FA in an ACV company 4-stage distillation system (Rutkowski *et al.*, 1978)

Symbol of acid	Content (% w/w)						
	in raw material	in fore-runnings	in fractions				
			C_{16}	C_{18}	C_{20}	C_{22} main	C_{22} side
$C_{16:1}$	0.3	2.2	11.4	—	—	—	—
C_{16}	3.2	12.2	79.3	1.4	—	0.2	—
$C_{18:1}$	12.0	20.9	4.0	39.7	0.7	0.1	—
$C_{18:2}$	8.6	19.4	3.1	28.4	0.2	—	—
$C_{18:3}$	8.5	22.3	2.2	28.1	—	—	—
C_{18}	1.1	1.4	—	1.2	2.1	0.9	—
$C_{20:1}$	9.6	5.8	—	1.1	96.0	1.7	1.6
$C_{22:1}$	56.7	15.8	—	0.1	1.0	91.7	98.4
Yield (%)	100	0.2	3.2	27.3	7.9	51.4	1.2

Table 1.22 Conditions and results of obtaining palmitic acid from fatty acid fraction of cotton-seed oil in a two-column system (Kločko and Vetkazova, 1976)

Conditions and results	I column	II column
Temperature (°C)		
head	145	215
middle	220	225
bottom	240	230
Pressure (hPa)		
top	9	10
bottom	45	28
Amount of distillate (%)	8	80
FA composition (% w/w)		
C_{12}	3.3	—
C_{14}	31.9	0.8
C_{16}	64.0	98.4
C_{18}	1.8	0.8

1.2.4.2 Crystallization

Crystallization from solvents is a method of low temperature fractionation, hence under conditions protecting against the formation of by-products. The method allows obtaining high purity FA or their fractions differing in solubility in solvents, most often polar ones. Methanol (Emersol process), acetone (Armour-Texaco process), hexane (CMB process, Italy), methyl ethyl ketone or methyl formate (Haraldsson, 1984) are most often used as solvents in industrial processes. Crystallization from solvents is used most often for the separation of acids of animal fats into so-called olein (liquid fraction) and stearin (solid fraction), as well as vegetable acids — from high-erucic rapeseed oil — into liquid C_{18} fraction and solid fraction of erucic acid.

In the Emersol process the FA in the amount of 30% are dissolved in 90% methanol and cooled down to 15°C. The precipitate is filtered off and washed with methanol. The solvent is removed from the precipitate and the filtrate by distillation. It is possible to obtain stearin of IV equal to 8 and a congeal point of 54°C and olein of IV equal to 93 and c.p. 3°C using this method for tallow FA (Haraldsson, 1984).

High-erucic rapeseed oil fatty acids are crystallized several times from acetone at –12°C (Fig. 1.41). Erucic acid of 99% purity is obtained (Rutkowski and Szczepańska, 1971).

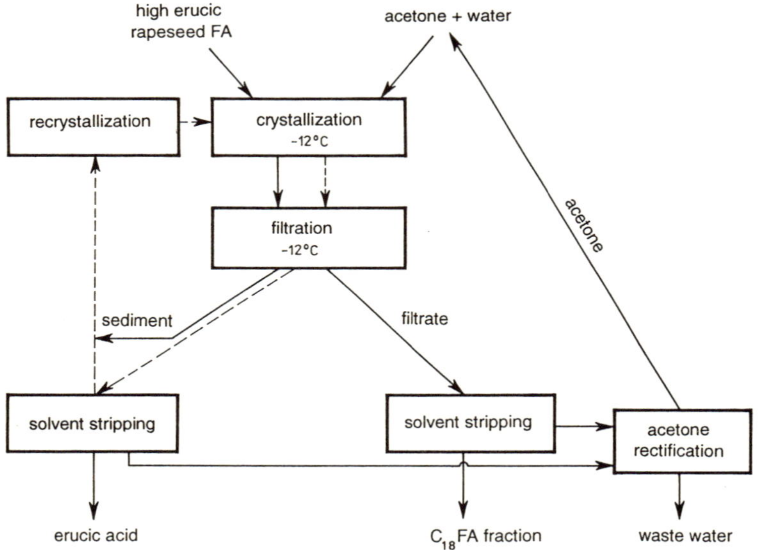

Fig. 1.41 Fractionation of high-erucic rapeseed FA (Rutkowski and Szczepańska, 1971).

Cotton-seed oil acids are fractionated into a solid and a liquid fraction by multiple crystallization from 33% solutions in gasoline at 0.7°C (Nabiev and Isaev, 1978). Much attention is paid in the processes of fractionation by crystallization from solvents to the recirculation of the solvent, methods and costs of its removal from final products, as well as protection of the environment by reduction of losses of the usually volatile solvent.

Stripping of 1 kg of acetone from a solution of a concentration higher than 10% requires 0.25 kg steam, and of 1 kg of methanol — 0.5 kg steam. The numbers are much higher when traces of the solvent are to be removed from FA fractions. Cooling 1 kg acetone vapours requires 8 kg of cooling water, while for 1 kg of methanol vapours 16 kg of water are needed (Grundman, 1984). Modern technological solutions are therefore based on a selection of the process conditions assuring proper fractionation quality at the lowest possible energy consumption. The costs of removal of traces of solvents from waste gases have been compared (Nitsche, 1984) for several methods, viz.:

— indirect condensation at a low temperature, depending on the amount of the solvent necessary to saturate the gas at a given temperature,
— absorption method,
— adsorption method,
— direct condensation.

1.2.4.3 Fractionation by hydrophilization

Fatty acids can be fractionated into a solid and a liquid fraction using aqueous solutions containing surfactants as wetting agents (Lanza, Henkel or Lipofrac process). A partly crystallized FA mixture is mixed with aqueous surfactant solution and is separated subsequently in a separator into a liquid fraction and aqueous suspension of wetted solid FA crystals, or using filters — into a solid fraction precipitate and a dispersive system of the liquid fraction in water (Szczepańska *et al.*, 1982; Fig. 1.42). The kind of the surfactant, its amount and concentraction, as well as the crystallization temperature, are chosen accordingly to the raw material being fractionated and the specifications for the products. Application of anionic surfactants is recommended. Aqueous surfactant solutions recirculate in the process, which is therefore almost waste-free.

The method is most often used for the production of olein from animal fat acids, although fractionation of other raw materials is also possible provided that a proper crystallization temperature is selected (Haraldsson,

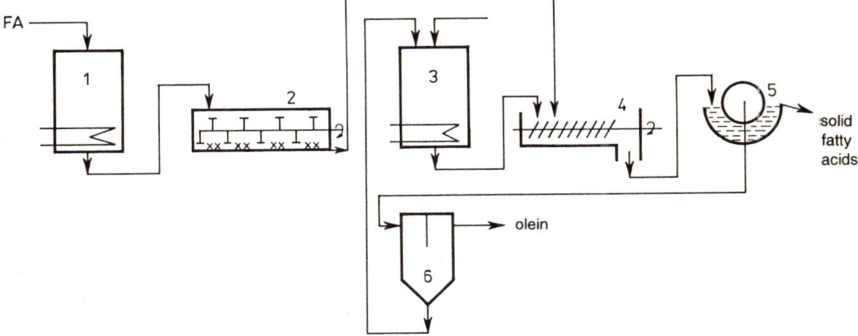

Fig. 1.42 FA fractionation by hydrophilization (Szczepańska *et al.*, 1982): 1—FA container, 2—scraping crystallizer, 3—wetting solution tank, 4—mixer, 5—vacuum filter, 6—separator.

Table 1.23 Results of fractionation of high-erucic rapeseed FA (Rutkowski *et al.*, 1978)

Symbol of acid	FA composition (% w/w)		
	raw material	erucic fraction	liquid fraction (by-product)
C_{16}	3.7	1.7	4.1
C_{18}	1.2	0.7	0.6
$C_{16:1}$	0.5	0.1	0.0
$C_{18:1}$	12.2	4.5	16.6
$C_{20:1}$	9.1	5.0	11.1
$C_{22:1}$	52.4	81.2	40.4
$C_{18:2}$	13.1	4.1	17.9
$C_{18:3}$	7.8	2.7	9.3
Yield (%)	100	35.1	52.7

1984). Two-stage fractionation at various temperatures is also carried out. About 80% erucic acid can be obtained by this method from high-erucic rapeseed oil FA (Table 1.23).

1.2.5 Hydrogenation of fatty acids

SFA, DFA, and even aqueous soapstock solutions from fat refining are subjected to hydrogenation (Abrachimov *et al.*, 1981). Hydrogenation of FA differs from hydrogenation of fats first of all in:

- —higher hydrogen pressure (2–3 MPa),
- —higher catalyst consumption, hence higher amount of a waste — the spent catalyst,
- —different requirements, since in hydrogenation of fats the selectivity of the process plays an important role, while in FA hydrogenation the main role is played by the reaction rate (the IV of the final product is reduced almost to zero).

The last factor leads to the application of continuous systems using suspended or stationary catalyst. Wastes amounting to 1.5–3.0% of the mass of the hydrogenated FA are obtained when using the traditional nickel catalyst containing 20–50% of Ni on a support (depending on the purity of the raw material) in an amount of 0.1 to 1% in terms of Ni. The amount of the waste depends on the method of catalyst separation. The FA hydrogenation catalyst is selected taking into account its activity, resistance to poisoning and stability in the FA medium.

Due to the presence of reactive carboxyl group, FA readily form so-called nickel soaps. These soaps — deposited on the catalyst surface — inhibit hydrogenation (Zschau, 1979). The phenomenon can be prevented when the catalyst is added to FA saturated with hydrogen and when the process is carried out under a pressure higher than 1.5 MPa. The effect of pressure on the amount of nickel soaps formed in hydrogenation of vegetable DFA using 0.5% Ni as a catalyst is the following (Kločko and Dorofeeva, 1972):

Pressure (MPa)	Nickel soap content (mg/kg)
0.5	155
1.5	45
2.0	30
3.0	20

The presence of sulphur, phosphorus, chlorine and nitrogen compounds in non-edible raw materials is the next reason for increased catalyst consumption, hence the increased amount of waste formed during FA hydrogenation. The effect of these compounds on catalyst consumption in FA hydrogenation is illustrated in Fig. 1.43.

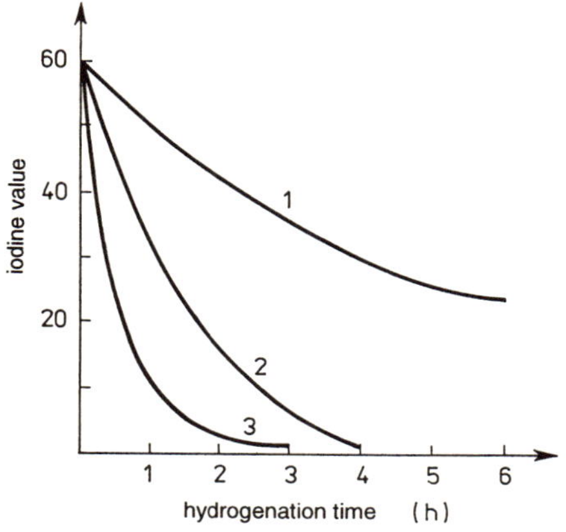

Fig. 1.43 The effect of purity of fatty acids on hydrogenation in the presence of nickel catalyst (Klimmek, 1984): 1—SFA, 0.06% w/w Ni, 2—SFA, 0.24% w/w Ni, 3—DFA, 0.06% w/w Ni.

Klimmek (1984) divides the impurities contained in FA into:

—catalyst poisons,
—catalyst inhibitors,
—catalyst deactivators.

Catalyst poisons comprise compounds containing elements from the groups V, VI and VII of the periodic table of elements with a free electron pair (S and P). It has been established for example that spent catalysts contained 0.5–2.5% S and 0.5–2.0% P. Sulphur compounds can be introduced not only with FA, but also with hydrogen (hydrogen sulphide, carbon disulphide, sulphur dioxide).

Examination of the effect of the amount and kind of poisons on hydrogenation yield (Fig. 1.44) revealed the strongest effect of chlorine

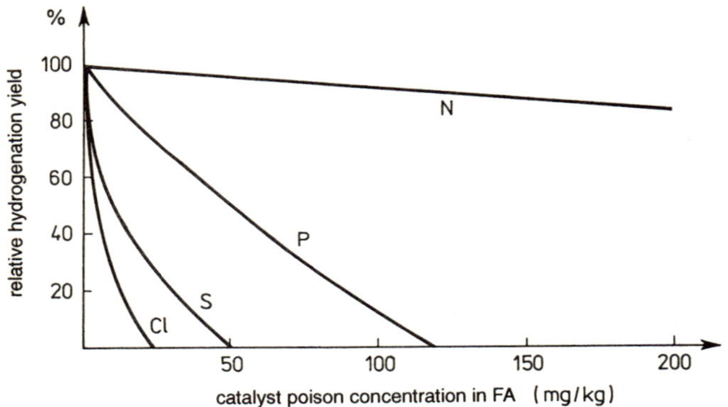

Fig. 1.44 The effect of catalyst poisons on the relative yield of hydrogenation (Klimmek, 1984).

compounds. The content of particular poisons in tallow SFA and the required increase in the amount of Ni catalyst (%) per 1 mg/kg of the poison are listed below (Klimmek, 1984):

	mg/kg	%
Cl	10–100	0.003
S	5–25	0.002
P	10–50	0.0006
N	200–600	0.0002

Nickel catalyst inhibitors comprise also the decomposition products: oxidized polymers, FA and alkali soaps. The inhibitory effect is due to physical blocking of the active sites of the catalyst. It has been established that 0.1% of oxidized FA requires an additional 0.003% of Ni, while 0.1% of sodium salts requires an additional 0.0002% Ni.

Catalyst deactivators are compounds reacting chemically with the catalyst, e.g. CO, being one of the strongest deactivators at temperatures lower than 160°C, and water. When the FA contain up to 1% of water, the additional consumption of the catalyst is equal to 0.008% Ni per 0.1% water.

The above references reveal that it is possible to decrease the consumption of catalyst significantly, and hence the amount of wastes formed during hydrogenation of fatty acids, by proper refining of the raw material. The most efficient, yet not always economical, is hydrogenation of DFA; hydrogenation of refined SFA with the aid of spent catalyst is cheaper, but less effective.

Modern trends in hydrogenation of FA are the following:

1. Application of very stable stationary catalysts. Aims at eliminating the cumbersome separation of the catalyst from the FA and at minimizing the amount of wastes formed in this process.
2. Application of continuous packed reactors of a special type, assuring largely increased contact surface area between the gaseous phase (hydrogen), liquid phase (FA), and solid phase (catalyst suspended in oil), which yields a high conversion degree in a very short time (Ilsemann and Mukherjee, 1978). Apart from solid wastes, in hydrogenation of both fats and fatty acids a gaseous waste also occurs, i.e. waste hydrogen. The hydrogen leaving the equipment contains water vapour and various gaseous and condensed components, e.g. CO_2, low molecular FA, acrolein, ketones, aldehydes, etc. One cubic meter of H_2 may contain up to 30 g of liquid fatty substances, 40–42 g of water and ca. 2% of the decomposition products (Kopylenko *et al.*, 1979).

Prior to its release to the atmosphere or to recycling the hydrogen must be purified. Purification of hydrogen on an industrial scale without the use of water condensers, hence without the formation of additional wastes, can be carried out in the following way (Prokopenko, 1978):

—hydrogen at 150°C is passed through two drip tubes and two dry condensers, where the liquid wastes are separated, the gas is cooled, and further condensation of the FA organic decomposition products occurs,
—hydrogen at 25–60°C is directed to the intertubular space of water cooled surface condensers, where part of the water vapour and the organic compounds are separated, the condensate being collected at the bottom of the condenser,
—the remaining water vapour is condensed after the complete decompression of the hydrogen in the next trap.

1.2.6 Obtaining of fatty alcohols

Fatty alcohols can be obtained from FA or their esters through hydrogen reduction. So-called alcohols of natural origin are obtained. Apart from them, alcohols occurring in nature (e.g. in some marine mammals) and synthetic alcohols of petrochemical origin are also known. Among these three groups the production of synthetic alcohols is the greatest. In 1983 it was equal in the United States to 400,000 t, in Europe to 130,000 t, and in Asia to 50,000 t, while that of fatty alcohols of natural origin was equal to 130,000 t, 220,000 t and 60,000 t at the above three continents, respectively (Kreutzer, 1984). The highest demand occurs for alcohols of a medium number of carbon atoms in the chain, so-called lauryl alcohols (ca. 57%); the rest consists of so-called tallow or detergent alcohols, saturated and unsaturated (Richtler and Knaut, 1984). Production of fatty alcohols plays an important role in non-edible fat processing, since they

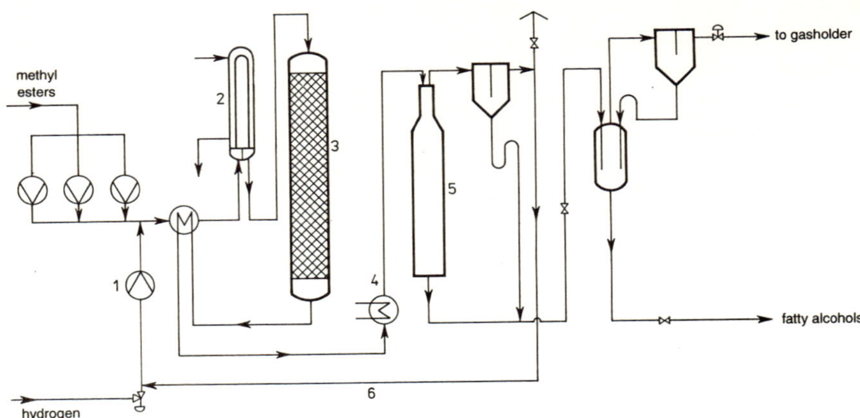

Fig. 1.45 System for methyl esters reduction using stationary catalyst (Kreutzer, 1984): 1—pump, 2—heater, 3—reactor, 4—cooler, 5—separator, 6—circulating gas, 7—flash tank.

constitute an intermediate product utilized mainly for the production of sulphated, sulphonated and oxyethylated derivatives, being readily biodegradable surfactants. These compounds are used for the production of washing agents and emulsifiers.

Fatty alcohols of a natural origin are obtained in the following ways:

— from glycerides or FA methyl esters by high-pressure reduction in the presence of suspended or stationary heterogeneous catalyst (Fig. 1.45), at a temperature of 250–320°C and under a pressure of 25–30 MPa,
— from FA by high-pressure hydrogen reduction, the acids being introduced to alcohols, thus occurring in the reaction medium as esters (Kreutzer, 1984) (Fig. 1.46). The reduction conditions are similar to those for esters.

The following compounds are used as suspended catalysts in the production of saturated alcohols: copper chromate, copper and nickel oxides, nickel or copper carbonates (Monick, 1979), copper chromate and cadmium, zinc and chromium (Zschau, 1979), zinc and copper oxides (Kreutzer, 1984). Copper, aluminium, zinc or chromium oxides (Porybleva and Zaiceva, 1976), as well as chromium-aluminium catalyst (Gorbačeva *et al.*, 1978b) are used as stationary catalysts. In the production of unsaturated alcohols catalysts selective towards the unsaturated bonds, e.g. zinc chromate, are used (Hinze, 1984).

Depending on the kind of raw material used the following by-products are obtained: post-reaction water from FA, methanol from methyl esters, or glycerine decomposition products, viz. a mixture of propylene glycols and propyl alcohol, from glycerides. Spent catalyst and waste hydrogen are the wastes in the production of alcohols. After the reduction the alcohols are purified by vacuum distillation, yielding an additional waste, the distillation residue.

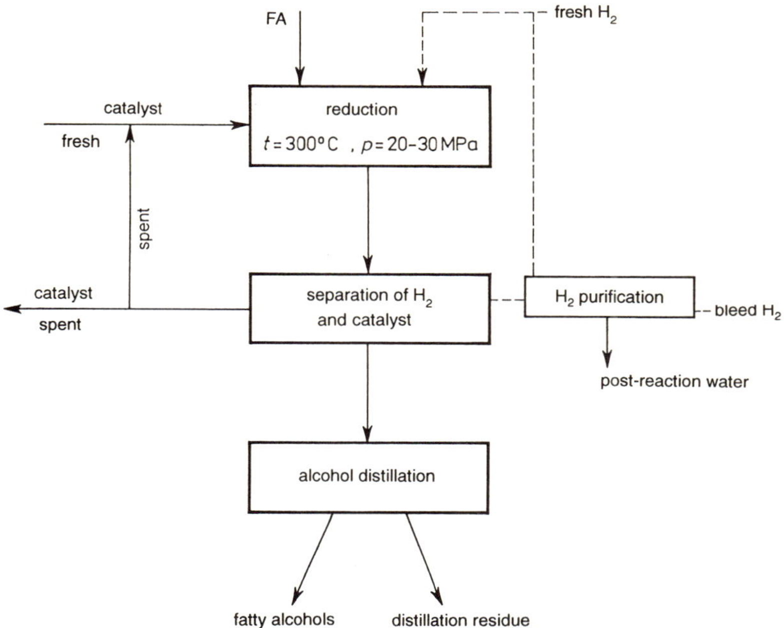

Fig. 1.46 Schematic diagram of the process of obtaining fatty alcohols of natural origin from FA, in the presence of suspended catalyst.

The amount of spent catalyst depends on the process applied and the purity of the substrates. The consumption of catalyst is reduced when the process is carried out at lower temperatures. The consumption of the suspended catalysts is 1–4% with respect to the raw material, while that of stationary catalysts is reduced down to 0.2–0.3% (Kreutzer, 1984). Stationary catalyst life can be prolonged when raw materials of high purity are used. Sulphur, halogens and phosphorus are the catalyst poisons. The content of each of these elements does not exceed 10 mg/kg. Spent catalyst from the process where it is suspended contains ca. 20% alcohols.

When methanol is the by-product, it can be distilled off and directed to methanolysis.

Raw fatty alcohols contain impurities, whose amount and kind depend on the process conditions. They are esters (2–5%), hydrocarbons (2–3%) and other decomposition products, as well as inorganic contaminants from the catalyst.

The alcohols are refined by distillation, sometimes preceded by alkali saponification of the esters to soaps (Kreutzer, 1984). The distillation is carried out under a reduced pressure, the temperature depending on the kind of alcohol. If the esters are not decomposed by saponification before the distillation, the amount of the distillation residue can reach 10% of the mass of the raw alcohols.

1.2.7 Saponification of fats and fatty acids

Production of soaps is one of the oldest domains of non-edible fat processing. Sodium or potassium soaps are obtained by alkali treatment of fats, fatty acidsor their mixtures. Apart from the traditional batch process of saponification of neutral fats or fatty acids in boiling kettles, the continuous processes are getting increasingly popular. Continuous production of soaps from fatty acids requires less complicated equipment than that for saponification of neutral fats.

The saponification process can be carried out at a temperature of ca. 100°C or more, under pressure. Sodium or potassium soaps containing ca. 62% FA are obtained in these processes. In the case of the production of toilet soap it is subjected to further treatment, including drying, chipping, mechanical treatment, etc. (depending on the kind of final product). A new kind of raw material for the production of soaps will probably be introduced in the future, viz. methyl or ethyl esters (Sonntag, 1981). This last method enables the production of high-grade soaps, having at the same time many other advantages, like better utilization of glycerine, which is separated at the stage of methanolysis (alcoholysis), and reduction of wastes. Soap lye is obtained as a by-product in the production of sodium or potassium soaps from neutral fats. Soap lye is an aqueous solution (ca. 10%) of glycerine, alcohols and salts. It contains organic impurities, like FA and their salts, as well as nitrogen and phosphorus compounds originating from the raw material impurities. Liquid alkaline wastes are the waste in the production of soaps.

1.2.8 Processing of specific non-edible fats

In the processing of non-edible fats there is a group of commonly processed raw materials, whose specific composition requires different processing methods. They are both waste materials — wool wax or tall oil, and natural oils, mainly castor oil. The specific feature of wool wax is the composition of its alcohols (high-molecular, *n*-, iso-, and ante-iso, as well as sterols) and fatty acids (high-molecular, *n*-, iso-, and hydroxyacids), of tall oil — the occurrence of large amounts of resin acids and sterols, while of castor oil — high content of 12-hydroxyoleic (ricinoleic) acid. Both the technologies of processing these raw materials, and the wastes, are different from those commonly encountered in the processing of non-edible fats.

1.2.8.1 Processing of wool wax

Wool wax, called suint, is a by-product of the textile industry, obtained during preparation of wool. Suint is a natural oiling agent, removed before spinning by washing the wool with soap or detergents, or by solvent extraction.

Depending on the kind, wool contains 5–30% of suint (Szczepańska and Szelejewska, 1979). Water from washing contains 1–3% of fatty matter. Suint is obtained from this water in various ways. The most common ones are the acidic, centrifugal and solvent methods. The yield of suint depends on the separation method and varies between 30 and 90% of the theoretical yield.

Centrifuging of suint from the water from washing allows ca. 60% of the initial content in the raw material to be obtained. The content of fatty matter in suint separated in this manner is equal to ca. 85%.

When detergents are used for washing the wool, part of them pass to the suint and hinder its further processing due to the formation of stable emulsions. Therefore from the viewpoint of further processing it is better to wash the wool with traditional soaps.

The yield of suint during degreasing of wool can be increased to ca. 95% by use of solvent extraction. Aqueous solutions of isopropyl alcohol, hexane, or both these solvents in turn (Vasileva *et al.*, 1974), as well as formic, acetic, propionic, oxalic or malonic acids (Teramoto *et al.*, 1973) are used as solvents. The concentration of suint in the waters from washing can be increased to 5–14% using membrane filtration (membrane pore diameter 1–1000 μm). Next, 0.3% (with respect to suint) of dioctyl sodium sulphosuccinate is added and the suint is separated by centrifuging (Lower and Ashworth, 1974). Suint yield in this method is increased to 95% with respect to the content in waters from washing.

Physical and chemical properties of suint depend on the sheep variety and region of breeding, the age of the sheep, wool storage conditions, as well as the method of wool degreasing (Table 1.24) (Vasileva *et al.*, 1974). Suint separated from wash waters contains less oxidation products, while that obtained by repeated extraction is better purified.

Table 1.24 Variations in the properties of wool wax depending on the method of obtaining (Vasileva *et al.*, 1974)

Property	Washing and extraction from the wash water	Ether extraction	Extraction with aqueous isopropanol and hexane
AV	2.4	12.1	7.0
SV	98	107.5	102
IV	26.5	25.1	27.1
m.p. (°C)	39.4	39.2	38.8
Colour*	0.190	0.189	0.080

* Of a 10% hexane solution relative to hexane, colorimetrically.

Suint is processed mainly into lanolin, i.e. refined, neutral wool wax. The refining process consists in alkali neutralization (using e.g. sodium carbonate) with simultaneous coagulation of the impurities, as well as bleaching, mainly chemical. The specific composition of lanolin alcohols and acids impart to lanolin and its derivatives emulsifying and coemulsifying properties. Due to this, all the operations carried out in aqueous media may be accompanied by the formation of stable emulsions. Water-alcohol alkali solutions can be used for neutralization. Isopropanol or amyl alcohol are used for this purpose. The yield of refined lanolin depends on the quality of the raw material and varies within the 70–93% limits. The yield is reduced to 60–65% when the neutralization is carried out in benzene solutions, since this solvent stabilizes emulsions (Busujuščij *et al.*, 1970). The by-products of alkali neutralization are aqueous solutions of lanolin soaps, containing emulsified suint and part of the non-fatty impurities.

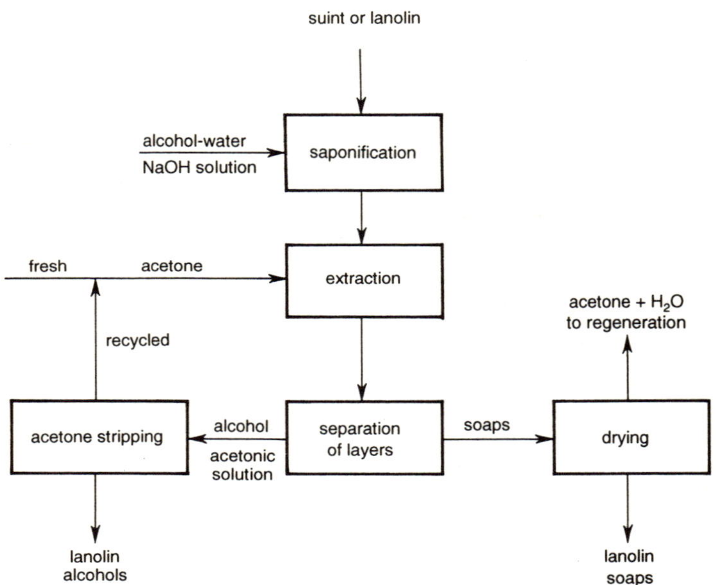

Fig. 1.47 Schematic diagram of the process of separation of lanolin alcohols.

Lanolin alcohols are obtained from suint or lanolin by saponification and extraction of alcohols (Fig. 1.47). Saponification with concentrated alkali solutions is carried out in the presence of isopropyl alcohol. The alcohols are extracted subsequently from the sodium or calcium soaps with acetone. A by-product is formed in this process — solid sodium or calcium soaps of lanolin acids in an amount of 50–60% relative to the raw material. A Japanese patent (Senda *et al.*, 1975) recommends pressure saponification at a temp. of 180°C and extraction with a mixture of ethyl acetate, water and $CaCl_2{\cdot}2H_2O$ in the proportion 400:120:16. Alcohols are separated from the ethyl acetate layer, while lanolin acids are obtained from the calcium soaps layer after acidification.

1.2.8.2 *Castor oil processing*

Castor oil is an unsual natural raw material, since it contains mainly hydroxyoleic (ricinoleic) acid glycerides, due to which it can be processed into special chemicals, like undecylenic acid, oenanthic aldehyde, 12-hydroxystearic acid, sebacic acid or dehydrated castor oil, i.e. glycerides of doubly unsaturated acid. Apart from the main products, a number of by-products and wastes are obtained in the above processes.

Undecylenic acid and oenanthic aldehyde are obtained by pyrolysis of castor oil or its methyl esters. The process is carried out at a high temperature of 450–700°C under a reduced pressure (300–400 hPa) within a few seconds. Oenanthal and undecylenic acid are formed as the main products of thermal decomposition of the oil, the yield being equal to 66–85.4% and 59.3–66.5%, respectively, and the purity exceeding 80%. Products of dehydration and oxidation of glycerine to acrolein are the volatile by-products.

Pyrolysis of methyl esters of ricinus acids, obtained by methanolysis of the oil, enables undecylenic acid methyl esters to be obtained with a yield higher by ca. 19%. Pyrolysis products are refined by distillation, and then rectification (Fig. 1.48). The following by-products and wastes are formed in this process (approximate percentage with respect to ricinus acids methyl esters):

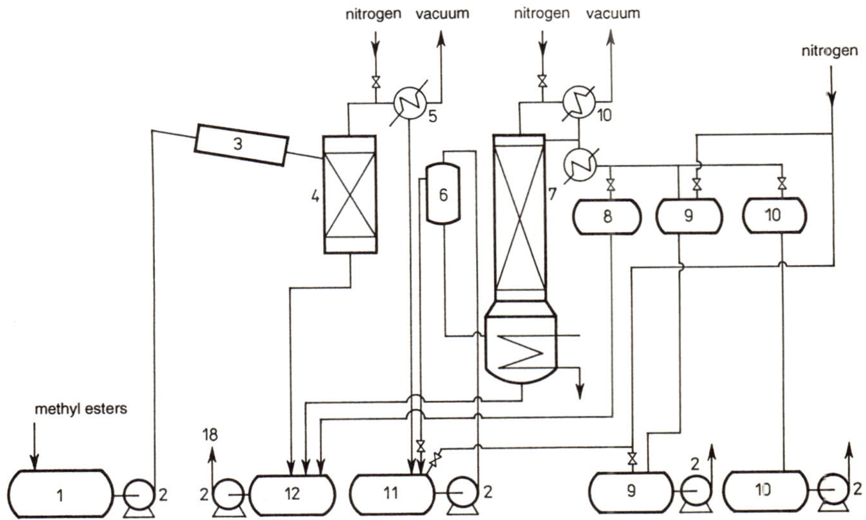

Fig. 1.48 System for obtaining oenanthal (Walisiewicz-Niedbalska and Witwicka, 1977): 1—methyl esters tank, 2—pump, 3—pyrolyser, 4—distillation column, 5—condenser, 6—notch tank, 7—rectifying column, 8—forerunnings and intermediate cuts tank, 9—oenanthal tank, 10—methyl undecylenate tank, 11—crude distillate tank, 12—waste tank.

simple distillation residue	10
rectification residue	18
forerunnings	8
intermediate fractions	6
volatiles	8

Forerunnings contain ca. 25% oenanthal, interfractions — ca. 10% oenanthal and 15% methyl undecylenate. The volatiles are mainly the products of fatty matter thermal decomposition, viz. CO, CO_2, ethane, ethylene, propane, propylene, etc. Residue I contains polymerization products and unreacted raw material, while residue II contains mainly methyl esters of saturated FA (Walisiewicz-Niedbalska and Witwicka, 1977).

12-Hydroxystearic acid is obtained from castor oil by hydrogenation, splitting and distillation under the conditions preservative for the hydroxyl group. The wastes formed are quantitatively similar to wastes from analogous processing of other raw materials (Sections 1.2.2, 1.2.3 and 1.2.5).

Sebacic acid (dicarboxyl C_{10}) is obtained by thermal decomposition (250–275°C) of ricinoleic acid in alkaline medium. Castor oil is introduced to concentrated KOH or NaOH solution being in a 100% molar excess. The octanoic alcohol formed is separated from disodium sebacate by distillation, extraction or distillation under a reduced pressure (b.p. of sebacic acid under a pressure of 3 hPa is equal to 143°C). Sulphuric acid is liberated in the process (Walisiewicz-Niedbalska and Witwicka, unpublished; Kadesh, 1979).

Industrial production of sebacic acid in the USA and Japan is based on oxidation of the oil in an alkaline medium at a temperature of 180–270°C in order to obtain disodium sebacate and octanoic alcohol. The soaps are decomposed by dilution and acidification to pH ca. 6. At this pH all the soaps of monocarboxylic acids are decomposed to FA insoluble in water, while the disodium sebacate forms monosodium sebacate, still soluble in water. After phase separation the aqueous layer is again acidified to pH 2, and the precipitated sebacic acid is separated by filtration (Johnson, 1984). Large amounts of liquid wastes are formed in this process.

Castor oil after dehydration becomes a mixture of glycerides of diunsaturated C_{18} acids. If the dienes are isomerized to give conjugation, a drying oil of properties similar to those of tung oil is obtained. Dehydration is carried out at 230–280°C in the presence of a suitable catalyst, e.g. a mixture of phosphoric acid, sulphuric acid, active aluminium or a mixture of $NaHSO_4$ and $NaHSO_3$ (Niedbalska and Witwicka, 1977). At the same time partial coupling of the unsaturated bonds takes place to an extent dependent on the kind of the catalyst. Application of 0.1% of concentrated H_2SO_4 or 0.4–0.5% of $(NH_4)_2SO_4$ at 260°C under a pressure of 20 hPa yields 20–30% conjugated bonds (Smirnov, 1972), while utilization of $NaHSO_4$ at 200–220°C under a pressure of 20 hPa allows obtaining ca. 50% conjugated dienes (Bhownick and Sarma, 1977).

Depending on the kind of the catalyst used either a solid waste (aluminium catalysts), or liquid wastes from neutralization and washing of alkaline or acidic catalysts are formed in this process. Dehydrated castor oil is used after polymerization for the production of oil varnish and oil paints.

1.2.8.3 Processing of tall oil

Apart from the basic substances of carbohydrate (cellulose) and aromatic type (lignin), wood contains also fats, waxes and resins. The amount of these components is low and depends on the kind of wood. It is accepted that the content of fatty and resin acids in ground wood is ca. 6%. In the production of sulphate pulp, based on boiling of wood in an alkaline solution of sodium hydroxide, sulphide and carbonate, so-called sulphate soaps dissolved in the waste lye are formed from fats, waxes and resins present in the wood. After concentration of the lyes the soaps are separated and acid hydrolyzed in order to obtain an oily product called crude tall oil. The yield of tall oil is estimated to be 53% with respect to the initial content in the raw material, while the proportion of fatty and resin acids in the crude oil is estimated at 76%. According to Logan (1979) the yield of crude tall oil relative to the mass of the wood is only 1–2%. Kelly (1980) anticipates that an increase in the production of tall oil will be achieved mainly by improving the yield. Table 1.25 lists the amounts of crude tall oil obtained from one ton of paper pulp in various countries. The yield of the oil depends on the variety, soil and climatic conditions of tree growth, their age, time of felling, duration and manner of wood storage, as well as the way of obtaining the oil from the wood.

The name "tall oil" is customary, since this product does not contain glycerides, but is a mixture of fatty and resin acids and unsaponifiable matter. It contains additionally various impurities. Crude tall oil is dark, due mainly to the presence of oxidized compounds, and has an unpleasant smell from sulphur compounds. The composition of the oil, like its yield, depends on the variety of the trees, region of felling and storage time,

Table 1.25 Yield of tall oil in some countries of the world in 1978 and 1980 (Surewicz and Surma-Ślusarska, 1983)

Country	Yield in kg/t of cellulose pulp	
	1978	1980
USA	22.4	22.7
Canada	7.6	7.9
Sweden	29.7	26.3
Finland	28.0	30.0
USSR	36.7	no data
Poland	38.9	36.0

the main role being played by the kind and variety of wood and the conditions of its growth.

World production of tall oil in 1980 exceeded 1.2 million ton annually. The main producers are USA, scandinavian countries, the USSR and Canada (Surewicz and Surma-Ślusarska, 1983). Applications of crude tall oil, being a waste raw material, are restricted. The known refining methods, based on action of H_2SO_4 on tall oil solution in petrol, action of hydrochloric or phosphoric acid, bleaching with chemical agents, e.g. H_2O_2, sodium chlorite or hyposulphate, or liquid-liquid extraction, have not found a widespread industrial use (Kocór and Kroszczyński, 1978). This is due to the poor effectiveness of the refining processes and simultaneous occurrence of decomposition processes of the basic components. On the other hand, industrial methods of tall oil refining by vacuum distillation, mainly coupled with rectification, have been developed. In 1980 the capacity of tall oil extraction plants was similar to the world production of this oil, which proves that it is the most rational method of utilization of this raw material.

Distillation of tall oil aims at the separation of unsaponifiable matter in the distillation residue (non-volatile) and in the forerunnings (volatile) and at obtaining as distillate the mixture of fatty and resin acids, or at a separation of fatty acids from resin acids. The quantity and quality of the main products (FA and resin acids), as well as by-products (interfractions), distillation residue and forerunnings, depend on the design of the equipment and the parameters of the process. All the components of tall oil are par-

Table 1.26 Main changes of resin acids during distillation (Holmbom, 1978a)

Type of change	Reacting acids	Products	Occurrence in fractions
Isomerization of unsaturated bonds	pimaric, isopimaric, sandaracopimaric, palustric, neoabietic abietic	isomers	TOR, DTO
Disproportionation	abietadienic	dehydroabietic, dihydroabietic	TOR
Dehydrogenation	abietadienic dehydroabietic	dehydroabietic, didehydroabietic	TOR
Decarboxylation*	all acids	C_{19} hydrocarbons and HCOOH or CO_2	TLO
Dehydration*	all acids	resin acids anhydrides and H_2O	TOR
Polymerization*	all acids except aromatic	dimers, trimers, etc.	TOR, TOP

* Reactions causing losses of resin acids.

ticularly susceptible to dehydration, oxidation and polymerization (Table 1.26); moreover, tall oil contains large amounts of admixtures, like higher alcohols, aldehydes and ketones, contained in volatile and non-volatile UM. They react at elevated temperatures with fatty and resin acids, decreasing the yield of the distillate and deteriorating the colour and odour.

Industrial systems for tall oil fractionation differ significantly from those discussed in Section 1.2.4 in the case of FA distillation. The following fractions are obtained in a typical equipment:

—tall oil rosin (TOR) — resin acid fractions containing up to 5% FA,
—tall oil fatty acids (TOFA) — containing over 90% FA,
—distilled tall oil (DTO) — a mixture of resin acids and 65–90% FA,
—tall light oil (TLO) — forerunnings containing volatile UM, like sulphur compounds, aldehydes, ketones, alcohols, as well as FA below C_{18} (Stage, 1978),
—tall oil pitch (TOP) — distillation residue.

The distillation residue is separated from crude tall oil (CTO) after dehydration in the first stage of distillation. The vapours are separated in a column. Fractions from the bottom of the column are the TOR, those from the middle — a mixture of fatty and resin acids, while from the top — the TLO. The mixture of acids is separated after rectification into TOFA and DTO. Separation of TOP takes place at a temperature of 200–250°C, using live steam, under a pressure in the column head equal to ca. 1 hPa and the pressure drop in the column ca. 10 hPa. Redistillation and separation of TOFA from DTO takes place under similar conditions. A system for the continuous fractionation of tall oil is presented in Fig. 1.49.

Average yield of the main and by-products from an industrial system (Table 1.27) can vary significantly depending on the design of the equipment (Holmbom, 1978b). Modern designs recommend the application of thin-film evaporators, assuring short residence time of the raw material and the intermediate products at a high temperature, and packed columns of low resistance, assuring small pressure drop (Knoer, 1970). Application of such techniques reduces the amount of TOP to ca. 80%, increasing simultaneously the yield of TOFA and particularly TOR.

Table 1.27 Average yield and characteristics of main and by-products of tall oil fractionation (Knoer, 1970)

Fractions	Yield (%)	AV	SV	Resin acids (%)	UM (%)
Tall oil pitch	16–20	ca. 30	ca. 110	5	
Tall oil rosin	38–42	165–175	180–195	85–94	ca. 3
Tall oil FA	24	ca. 194		1	2
Distilled tall oil	7	ca. 187		25	
Tall light oil	9	ca. 110			
Losses	2				

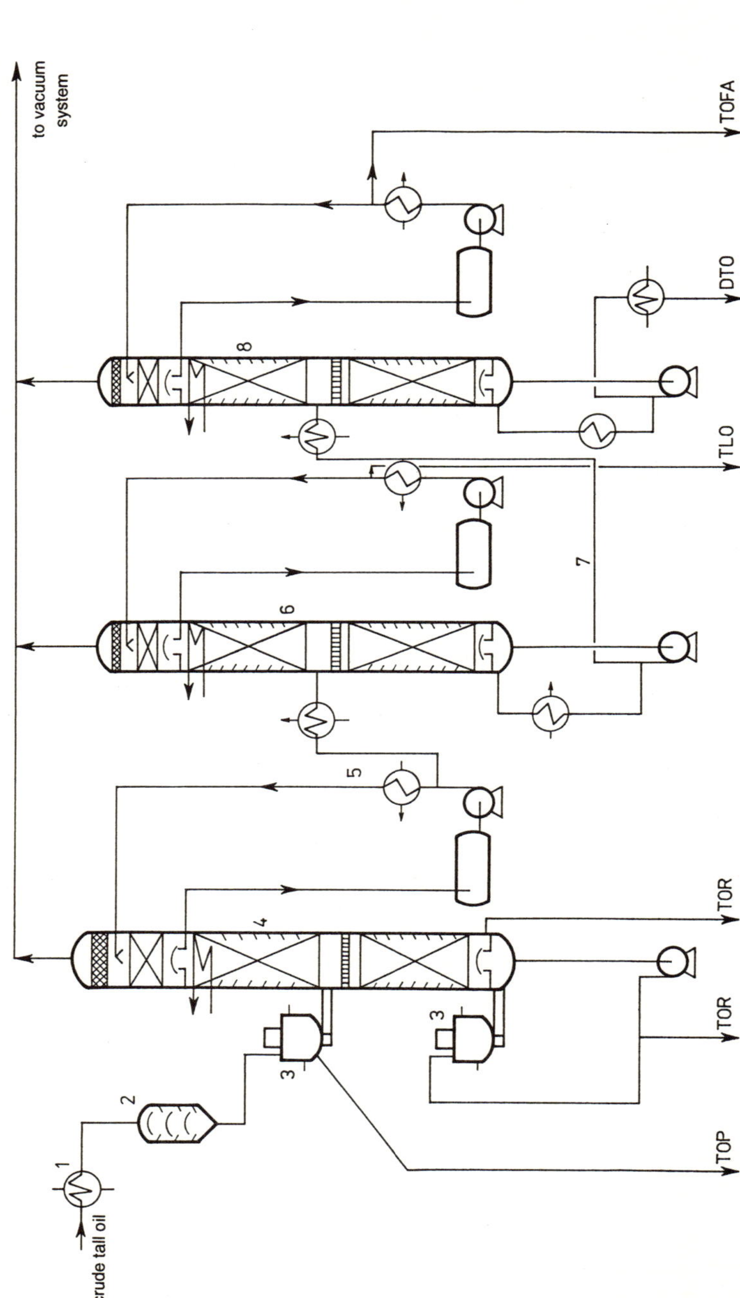

Fig. 1.49 Tall oil distillation system (Kehse, 1976): 1—heat exchanger, 2—drier, 3—evaporator, 4—column I, 5—tank for intermediate cut I, 6—column II, 7—tank for intermediate cut II, 8—column III.

1.2.9 Processing of other fatty derivatives

On a world scale the non-edible fats constitute a significant source of raw materials for the production of a wide variety of derivatives called oleochemicals. Although the technologies of obtaining them are considered to be fat processing, yet taking into account their chemical composition they are very diversified compounds.

The technologies of basic oleochemicals, viz. FA and fatty alcohols, have been discussed in the previous paragraphs. However, due to a rapid development of the production of various fatty derivatives it seems purposeful to briefly present some of the other processes, being the sources of various by-products and wastes in amounts significant to the natural environment.

1.2.9.1 Fatty acid methyl esters

FA methyl esters are introduced increasingly often as a raw material for the production of oleochemicals. Methyl esters can be obtained directly from non-edible fats and oils by alcoholysis. This process is less energy-consuming than splitting, and it additionally yields high-grade glycerine as the by-product. Fractionation of FA methyl esters is more effective than of fatty acids, because the esters have lower boiling points and the differences between the homologues are greater. Moreover, they are more thermally stable and do not form anhydrides, hence less wastes are formed during their processing. About 32% of fatty acids (excluding tall acids) was obtained in 1983 in the USA in the form of methyl esters (Sonntag, 1984). Methyl esters are used primarily as the intermediates in the synthesis of alcohols, amines, nitriles, ester dimers, etc. (Farris, 1979).

Natural oils can be esterified at a temp. of 50–70°C using an excess of methanol at atmospheric pressure. About 90% glycerine is a by-product

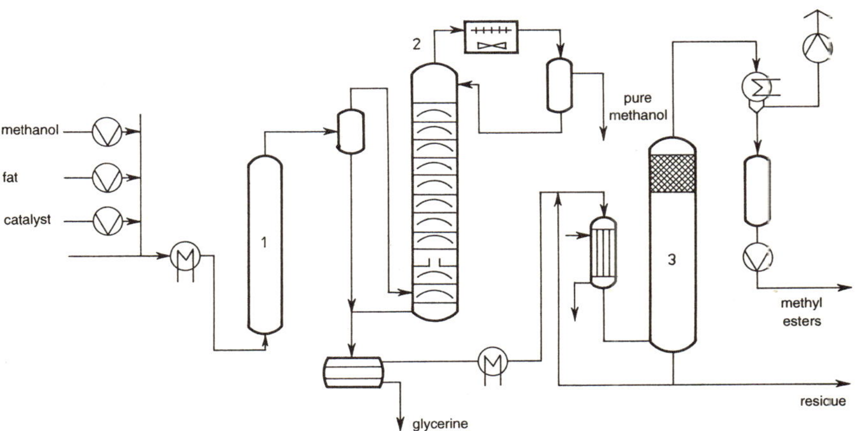

Fig 1.50 System for obtaining methyl esters from fats (Kreutzer, 1984): 1—reactor, 2—glycerine separation section, 3—distillation of methyl esters.

from this process. Unreacted methanol recirculates and must be rectified to remove water. The alcoholysis can also be carried out in the presence of an alkaline catalyst at a temperature reaching 240°C under a high pressure, with a few mole methanol excess. The pressure method permits the use of oils and fats of high acidity (Kreutzer, 1984).

Esterification of FA with methanol can be carried out in a system presented in Fig. 1.50. Heated FA with an alkaline catalyst are transferred to a plate column countercurrently to methanol introduced at the bottom of the column. The reaction is carried out under a pressure of ca. 1 MPa and at a temp. of 240°C, maintaining a 3–4 fold excess of methanol with respect to the FA. Methanol excess is carried away together with water from the top of the column (1:1 methanol-water mixture). Methyl esters of the AV lower than 0.5 are separated from the catalyst by distillation.

Esterification of FA with methanol can be also carried out in the presence of an acidic catalyst, e.g. H_2SO_4 or organic sulphoacids. The reaction takes place during the passage of the methanol vapours through a mixture of FA with the catalyst at ca. 110°C. Methanol vapours and water from the reaction are transfered to rectification and methanol is recycled. Using a 0.5% addition of 60% H_2SO_4 it is possible to reach ca. 97% conversion by this method, the AV being equal to 6. Using more concentrated H_2SO_4 it is possible to achieve higher conversion degree (AV ca. 2.5), but the unsaturated acids undergo partial oxidation and the IV for cotton seed FA decrease by ca. 15 units (Gorbačeva *et al.*, 1978a).

No by-products are formed during esterification.

1.2.9.2 Fatty acid monoglycerides

A technical product called monoglyceride is usually a mixture containing 40–48% mono-, 30–40% di- and 5–10% triglycerides. It contains additionally 0.2–9% FFA and 4–8% glycerine (Meffert, 1984). Monostearates and monooleates are most commonly produced. High grade monoglycerides are obtained by refining the technical product. Methods yielding medium-grade monoglycerides (ca. 60% monoglyceride content) are also known (Van Haften, 1979).

Monoglycerides are obtained by direct esterification of the chosen FA with glycerine or by glycerolysis of fats, oils or hydrogenated oils. Esterification is carried out at 150–260°C at a reduced pressure, without or with acidic or alkaline catalysts. Glycerolysis can be carried out without the catalysts, but the necessary temperature of ca. 280°C results in the formation of large amounts of decomposition products. Alkaline catalysts are most often used on the industrial scale, e.g. NaOH, KOH or $Ca(OH)_2$ in an amount of 0.05–0.2%. After completing the process the catalyst is neutralized with phosphoric acid, and the resulting sodium, potassium or calcium phosphates are removed by adsorption on bleaching earth (Sonntag, 1982).

Numerous investigations aiming at developing a method of direct production of high-grade monoglycerides have been carried out. However, none of the patented methods is ideal. Glycerolysis in emulsified systems can be mentioned here. The content of monoglycerides increases in this method by 10–30%, but the large amount of soaps is the disadvantage. The glycerolysis is carried out in the following solvents: dioxane, pyridine, phenols, cresols (ca. 80% product) (Sonntag, 1982). Esterification of oleic acid in dioxane in the presence of an ion exchanger with hydrophilic groups serving as a catalyst yields even up to 90% purity monooleate of a light colour (Černyševa and Šmidt, 1975). Glycerolysis of methyl esters carried out under a reduced pressure with an excess of glycerine and simultaneous removal of methanol enables increasing the monoglyceride content to 90% (Sonntag, 1982).

It is anticipated that enzymatic synthesis will be the future method of synthesis of partial glycerides (Hoq *et al.*, 1984). The process is carried out in a membrane reactor. FA react with glycerine, containing 3–4% of water and bacterial lipase. The reaction proceeds in a continuous way, and both the substrates circulate (Fig. 1.51). The product obtained contains sole-

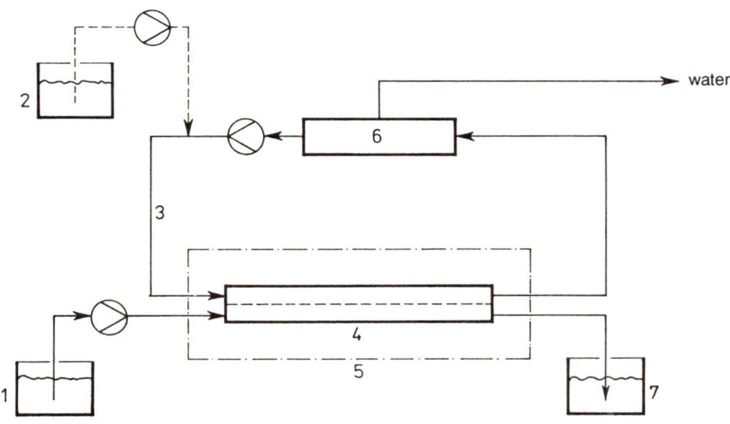

Fig. 1.51 Enzymatic esterification instrumentation (Hoq *et al.*, 1984): 1—FA tank, 2—glycerine tank, 3—glycerine, water and lipase mixture, 4—membrane bioreactor, 5—thermostated water bath, 6—deodorizer, 7—glycerine tank.

ly 1-mono- and 1,3-diglycerides (mole/mole) and is very pure — without any decomposition products or water. No wastes are formed in the process.

High-purity monoglycerides are obtained by molecular distillation (Fig. 1.52). The glycerolysis is carried out at a temperature of up to 235°C. Higher temperature causes dehydration of glycerin molecules and the formation of polyglycerols and acrolein. Polyglycerols have a b.p. similar to monoglycerides and are not separable from them by molecular distillation. Polyglycerol formation is promoted by excessively high concentration of alkaline catalysts — their concentration should therefore not exceed 0.1%. To prevent the displacement of the reaction equilibrium the catalyst and

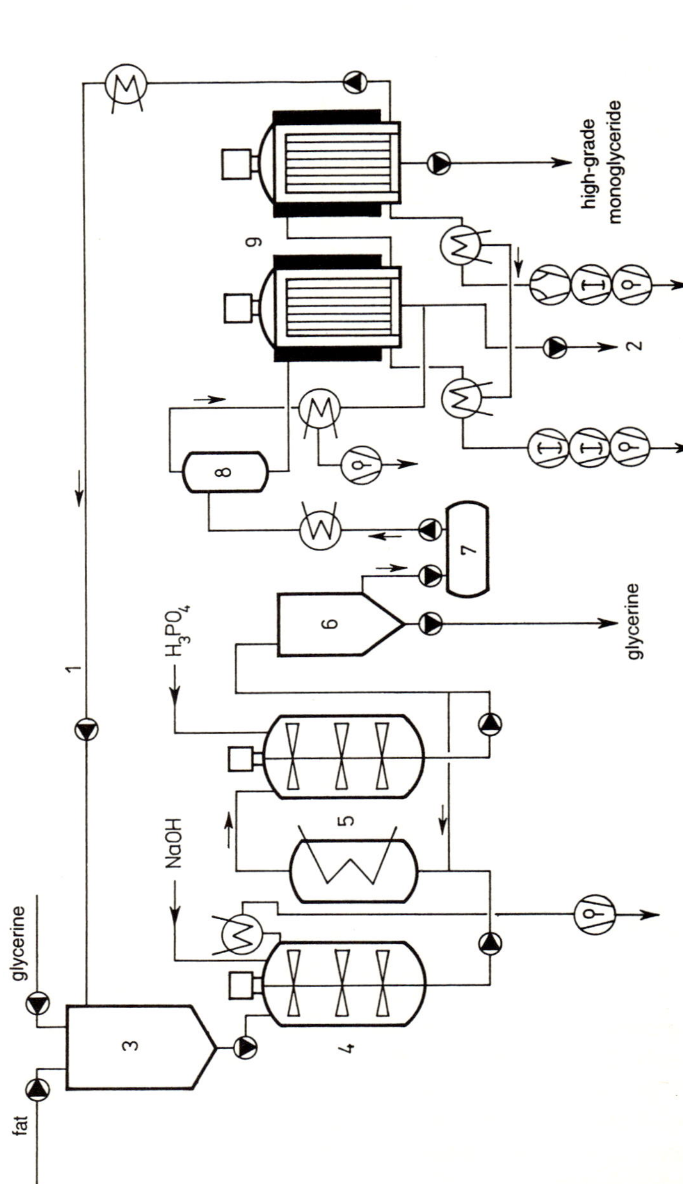

Fig. 1.52 Installation for obtaining high grade monoglycerides (Fischer, 1981): 1—diglyceride, 2—forerunnings, 3—tank, 4—reactor, 5—neutralizer, 6—separator, 7—tank, 8—evaporator, 9—short path evaporator.

glycerine are immediately removed from the reaction medium. The catalyst is removed by stoichiometric neutralization. Maintaining these conditions enables efficient distillation of monoglycerides and recycling of the by-products. A three-stage distillation refining is recommended, viz.:

1. Removal of water and of a part of glycerine at a temp. of 180°C and under a pressure of 2–4 hPa.
2. Removal of the remaining forerunnings at 190°C and under a pressure of 0.1 hPa.
3. Distillation of monoglycerides at 200°C and under a pressure of 0.05 hPa.

The residue and the forerunnings can be recirculated 5–8 times to glycerolysis in a batch process (Fischer, 1981).

Low, middle and high-grade monoglycerides are commonly used as emulsifiers for food products. They have also many technical applications.

1.2.9.3 Oxyethylated derivatives

Oxyethylated derivatives of acids or fatty alcohols are non-ionic detergents or emulsifiers. FA or fatty alcohol aliphatic chains constitute the hydrophobic moiety of the molecule, while the chain of ethylene oxide groups is responsible for the hydrophilic properties. These properties depend on the number of ethylene oxide groups attached. Owing to this, products completely insoluble or soluble in water can be obtained by this method. Oxyethylated FA are usually less soluble and have poorer foaming and wetting properties than the analogous derivatives of fatty alcohols.

Industrial method of obtaining oxyethylated derivatives is based on gradual addition of ethylene oxide to FA at 110–180°C and in the presence of 1–3 mole % of an alkaline catalyst (with respect to FA). Batch designs using pressure reactors with stirrers or circulating reactors are most often utilized. The reaction is carried out in an inert gas atmosphere. After completing the process, the products are cooled, the catalyst is neutralized, traces of ethylene oxide are removed and the neutralized catalyst is filtered off. The neutralized catalyst is the only waste from this process (Stockburger, 1979).

1.2.9.4 Sulphated and sulphonated fatty alcohols and acids

Sulphate derivatives of fatty alcohols, oxyethylated fatty alcohols or fatty acids are surfactants. Modern methods of production are based on the action of gaseous SO_3 on fatty alcohols or FA in apparatus of a special design, assuring rapid cooling of the reaction mixture flowing in the form of a thin film (Lanteri, 1978). The Sulphurex F system can be taken as an example of a modern solution (Fig. 1.53).

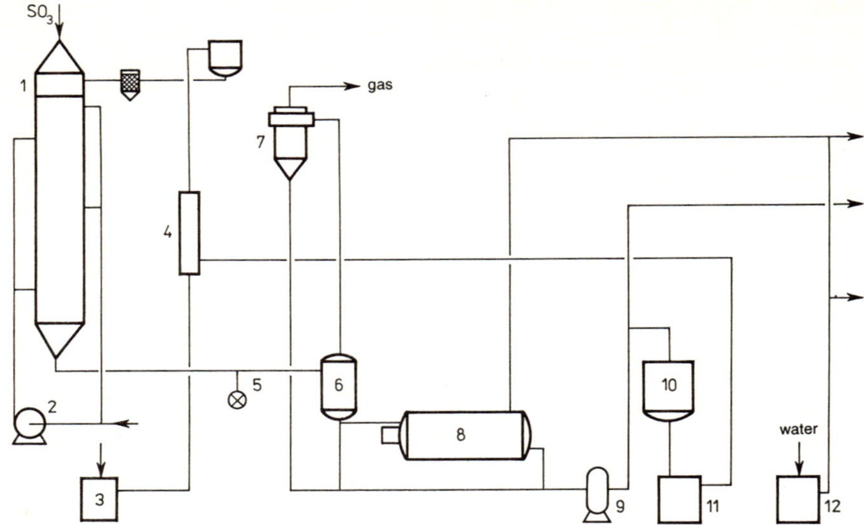

Fig. 1.53 Sulphurex F system for sulphation and sulphonation (Davidsohn and Moreth, 1985): 1—multi-tube reactor, 2—cooling water circulation pump, 3—raw material feeder, 4—static mixer, 5—sulphonation degree indicator, 6—liquid and gas separator, 7—outlet gas cyclone separator, 8—transition tank, 9—acid pump, 10—acid tank, 11—acid feeder, 12—water feeder.

The product obtained is neutralized. The only dangerous waste is the fumes.

1.2.9.5 Alkanolamides

Alkanolamides are the products of condensation of FA mainly with monoethanolamine and diethanolamine, called mono- or dialkanolamides, respectively. Alkanolamides have properties complementary to surfactants, since they increase their washing and foaming abilities. They also reveal protective, stabilizing and thickening properties. They are used as additives for washing powders, soaps, shampoos, as well as washing and cleaning fluids. Glycerides, FA or their methyl esters are condensed with mono- or diethanolamine. Condensation conditions depend on the type of substrate. Methyl esters condense more rapidly and at lower temperatures than FA (ca. 100°C and 170°C, respectively). The process can be catalyzed by e.g. 1% sodium methylate (Cahn, 1979). Apart from the main compound, the products obtained contain free amine used for the condensation and other impurities (Table 1.28). In the case of glyceride condensation glycerine is not recovered. Condensation of methyl esters yields hydrated methanol as a by-product. Following a rectification, it can be recycled to methanolysis of fats.

Table 1.28 Typical composition of diethanolamides (Farris, 1979)

Components	Content (%) in diethanolamide obtained from	
	fatty acids (2 : 1 amide)	methyl esters (1 : 1 amide)
Diethanolamide	55	90
Amido-amine	10	traces
Free diethanolamine	22	5
Amine soaps	10	traces
Amide esters	1	4
Water	2	traces
Methanol	—	0.2
Methyl esters	—	0.8

1.2.9.6 Fat amides

First-order amides of fatty acids ($RCONH_2$) are obtained in a reaction of FA with anhydrous ammonia. The reaction is carried out at 180–200°C under a pressure of a few MPa. The water formed is carried away from the reaction medium together with the unreacted ammonia. The ammonia is recovered from the waste condensate and recycled to the reaction. The reaction can be catalyzed by e.g. boric acid, Al_2O_3 or zinc compounds

The synthesis can be carried out using methyl esters as the substrate. In this case methanol is liberated in the reaction. It is next rectified and directed to methanolysis.

Fatty amides are used as assisting agents in the plastics industry, synthetic fibre industry, etc. (Reck, 1979).

1.2.9.7 Fatty nitriles

Fatty nitriles ($R—C{\equiv}N$) are obtained by ammonolysis of FA followed by dehydration. They are formed in a reaction of FA with ammonia at 260–280°C, or even 360°C, in the presence of a catalyst, e.g. ZnO (0.1–0.25%) (Hinze, 1984). The process can be carried out in a continuous manner. In such a case the FA vapours pass through a tube reactor containing a catalyst (e.g. Al_2O_3), countercurrently to ammonia vapours. Ammonia and water excess are collected at the top of the column. Ammonia is recovered in an ammonia absorber and recycled (Reck, 1979).

Direct synthesis from glycerides is proposed in modern procedures (Billenstein and Blaschke, 1984). Fats or oils react with ammonia at a temp. of 230–290°C at atmospheric pressure, in the presence of special catalysts (Fig. 1.54). Glycerine washed with water from the reaction product is the only by-product.

Fatty nitriles are produced mainly as a first stage of fatty amine synthesis. They are also used as improvers for special lubricants.

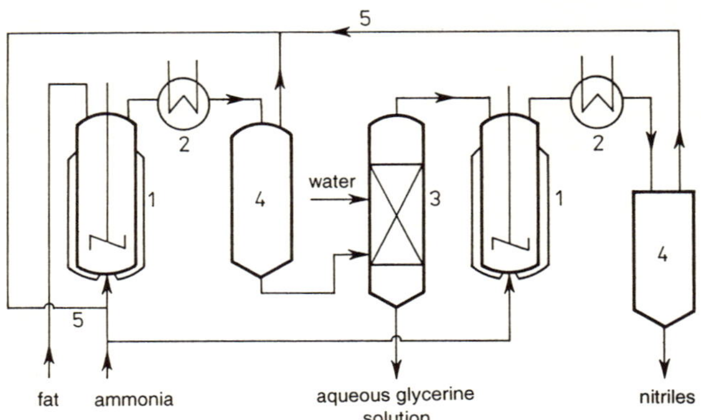

Fig. 1.54 Schematic diagram of the equipment for nitrile production (Billenstein and Blaschke, 1984): 1—reactor, 2—condenser, 3—extractor, 4—tank, 5—recirculating ammonia.

1.2.9.8 Fatty amines

Hydrogenation of FA nitriles yields fatty amines (RCH_2NH_2). The process takes place in the presence of ammonia vapours, within a 120–180°C temperature range, under a pressure of 3.5 MPa or more, using Raney Ni or Co as a catalyst (Reck, 1979; Billenstein and Blaschke, 1984). Spent catalyst is the waste. Primary fatty amines, in particular, find widespread application as additives for lubricants, corrosion inhibitors, fuel additives, flotation agents, anti-caking agents and intermediate products for further syntheses.

1.2.9.9 Acid or ester dimers

At high temperatures (200–240°C) and in the presence of a catalyst (mainly activated montmorillonite clays), the unsaturated fatty acids or their methyl esters form acid or ester dimers by combination of two molecules of unsaturated acids. The conversion degree is 40–80%, depending on the conditions of the process. Apart from the main product, higher oligomers and unreacted monomers of a constitution different from that of the raw material are also formed (Fig. 1.55).

The catalyst is separated by filtration, and the monomers are distilled off in thin-film evaporators under a pressure of 1–5 hPa and at 290–300°C (Szczepańska *et al.*, 1982b). Dimers and higher oligomers are contained in the so-called spent liquor. Dimers freed from oligomers are obtained by molecular distillation. The by-products of the process are distilled monomers (in the form of esters or saturated and unsaturated acids of odd and even carbon atom number, in *cis* and *trans* isomeric forms). The

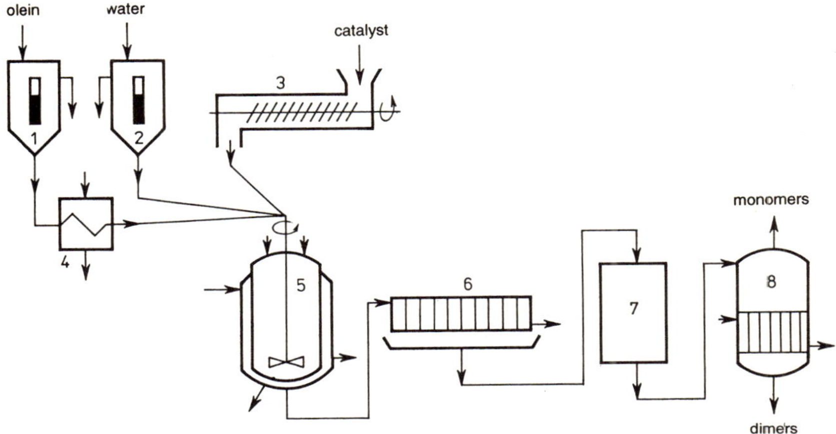

Fig. 1.55 System for obtaining acid dimers (Szczepańska *et al.*, 1982): 1—olein meter, 2—water meter, 3—catalyst feeder, 4—heater, 5—autoclave, 6—filter press, 7—tank, 8—evaporator.

amount of this product reaches 20–40 % with respect to the raw material used for the dimerization.

The waste is the catalytic clay, containing ca. 50% of partly polymerized fatty matter (ca. 3% of waste relative to the raw material). The fumes from the dimerization reactor contain volatile FA decomposition products and must be freed from organic compounds before releasing to the atmosphere.

In the USA the acid dimers are produced in significant amounts from tall oil FA and from technical oleic acid. The dimers are mainly used for the production of polyamide resins (Leonard, 1979).

1.2.9.10 Polymerized oils

Neutral oils of high unsaturated acids content are subjected to thermal oxidative polymerization (with air) in the presence of metal and FA salts, so-called siccatives. Salts of Co, Pb or Mn in an amount equal to 0.01–0.1% (in terms of the metal) are used as the siccatives. Air is passed through the oil heated to 80–160°C. This results in thickening of the oil due to the formation of polymers of various composition (Melnikov and Mnuchin, 1980). These oils are used for the production of oil varnishes.

No by-products are formed in this process, but the fumes are dangerous for the environment, since they contain volatile products of fatty matter decomposition.

1.2.9.11 Epoxidized oils

Oils containing polyunsaturated acids, e.g. soyabean oil, are oxidized with organic peracids formed *in situ*. The process is carried out at 50–70°C, using acetic acid and hydrogen peroxide, in the presence of an acidic catalyst, e.g. H_2SO_4 or Amberlite JR-120 ion exchanger (El-Shami *et al.*, 1985). Epoxidized oils are used mainly in the plastics industry and the preventive coatings industry.

1.3 OTHER INDUSTRIES

The world chemical industry is based to a large extent on non-edible wastes from the food industry. Vegetable post-refining acids are a significant source of raw materials. A large part of the non-edible raw materials discussed are used nowadays as feed additives, e.g. waste animal fats or poultry fats. The selection is becoming increasingly rigorous, which creates a prospect for the oleochemical industry of further deterioration of raw material quality. In numerous cases the interests of the producer and the recipient of waste fats are opposed, since the producers try to reduce the amount of waste and by-products or to improve their purity to a level required for feeds.

1.3.1 Meat processing

Raw tissue fats from slaughter animals are mainly consumed directly, e.g. back fat, or after rendering — lard, beef tallow. Table 1.29 lists the yield factors of various raw fats from the meat processing industry, and the way of their utilization (Urban, 1979).

Tissue animal fats can be melted down by various methods in meat processing plants or in technical rendering plants. Drippings of various quality are obtained depending on the kind of the raw material, its freshness, parameters and method of rendering. The following kinds of lard are distinguished in the USA, being the main producer of pork fat (Bernardini, 1973):

- neutral lard wet rendered from selected raw materials at low temperature, used solely for human consumption,
- dry rendered leaf fat,
- soft fat from the residue from the above two processes and from the remaining fat tissue, rendered under pressure at 140°C — used mainly for technical purposes.

Beef fat, called tallow, is obtained from adipose tissue surrounding stomach (omental fat) and intestines (capsular fat). High grade beef fats are used primarily for direct consumption, especially in various types of food concentrates. Lower grades are used for technical purposes.

Table 1.29 Crude fats from meat processing industry (Urban, 1979)

Name	Raw material	Per cent content of fatty matter in the raw material (%)	Yield of fat (%)	Fatty matter content in crude fat (%)	Yield of crude edible fat (%)	Yield of crude non-edible fat (%)
Perirenal and beef tissue tallow	beef carcasses	8.3	33.7	80	99. 0	1.0
Slaughter beef tallow	cattle and slaughter calves	3.2	87.5	80	5	94
Pork leaf fat	fatteners	2.8	93.0	85	99.0	1.0
Back fat	pork carcasses	8.0	81.0	90	99.0	1.0
Pork tissue fat	pork carcasses	16.0	16.3	85	100	—
Pork slaughter fats	fatteners	2.1	76.2	76	20	80

According to Maerker (1979), edible fats are classified mainly with respect to precisely defined conditions of production hygiene. In the United States ca. 8% of the tallows produced are classified as edible.

The characteristics of the fats obtained depend on numerous factors, including the rendering method: wet, dry, at a temperature higher or lower than 100°C, in a continuous or batch manner (Table 1.30). According to Maerker (1979), microwave heating will be applied in the rendering plants in future.

1.3.2 Feed processing

Utilization greases are a by-product obtained in utilization plants during processing of various raw materials (carrion, slaughter seizure and by-products, sludges, spoiled food products, technical cracklings, etc.) into feed meals.

Carrion is defined as the bodies of animals that died or were killed in order not to be consumed. Depending on the kind, it can be divided into horse and colt, beef and calf, pig, small animals' and poultry carrion. Meat-and-bone meals and non-edible fats are primarily produced from carrion. The value of carrion depends first of all on the age of the animal.

Slaughter seizures are the pathologically changed carcasses, parts of carcasses or organs, not allowed for consumption by proper regulations. Slaughter by-products are the animal foetus and inedible parts of animals, like reproductive organs, lips, etc.

Table 1.30 The effect of the rendering method on the technical properties of animal fats. Tests carried out on a mixture of 77% of pork intestine wastes and 23% of slaughter animal bones (Fajviševskij and Mginaridze, 1976)

Rendering method	AV	m.p. (°C)	FA c.p. (°C)	Hardness (g/cm)	IV	Phosphorus content (°C)	UM (%)
1 h at 90–95°C 1–1.5 h, p = 196–294 kPa, drying for 2–3 h, straining out, settling at 60–65°C	3.4	39.1	41.7	22.5	46	0.00–0.04	0.85
45–60 min, p = 196–294 kPa, drying for 2–2.5 h, straining out at 60–65°C	3.8	38.8	40.5	234	44	0.00	0.72
2–3 h, p = 196 kPa, boiling to a pulp in water, settling, bouillon decanted and settled	19.0	36.9	40.5	128	—	0.04–0.08	0.54
Degreasing in a press, 8 min at 90–95°C, settling of the fat	4.2	39.3	41.2	234	45.5	0.00	0.57
Degreasing in a press, 20 min at 90–95°C, settling of the fat	5.9	38.2	38.7	231	46.6	0.00	0.32

Sludges are formed during cleaning of intestines — they are their internal and external layers. The quality of sludges is very variable.

Technical cracklings are the non-edible residuals from rendering of fats. Depending on the rendering method they can be pressed or not pressed, autoclaved, or they can originate from the Tytan-type installation. The chemical composition of cracklings is presented in Table 1.31.

Table 1.31 Chemical composition of technical cracklings (%)

Kind of cracklings	Water	Fat	Protein	Ash
Pressed	35–50	20–35	19–34	2–4
Non-pressed	15–30	30–45	30–45	3–6
From Tytan installation	50–60	8–35	9–19	2–4

It is generally accepted (Urban, 1979) that the average fatty matter content in the utilization raw material is 20%, and that the yield of fat is equal to 60%. All the fat obtained is used for technical purposes.

The technologies of utilization grease and feed meal production must meet certain requirements. Due to the kind of the raw material, it is necessary to kill all the bacteria; the yield and kind of both the products (grease and meal) are also important.

Utilization greases are obtained in the following processes (Bernardini, 1973):

— comminution of the raw material and action of live steam at 50°C, followed by stirring and homogenization in a disintegrator, separation of the water-oil phase residue, heating of it and centrifuging of the fat (this method does not assure elimination of bacterial colonies),
— dissolution of the raw material for a few hours at a pressure of 2–3 MPa in an autoclave, followed by a reduction of the pressure and evaporation of a part of water down to 5–7%, and finally by separation of the fat from the solids in horizontal centrifuges,
— solvent extraction of the fragmented raw material and obtaining of fat from miscella. The meal is obtained after the removal of the solvent.

Production of utilization greases in the Hartman type equipment consists of the followings stages:

1. preparation of the raw material (skinning, partition, fragmentation, washing),
2. loading into a destructor, supplying of steam and boiling for 30 min at 130°C under a pressure of 0.2–0.3 kPa with simultaneous sterilization,
3. reduction of the pressure and drying under a reduced pressure at a temp. of 100°C, aiming at reducing the water content down to 6–11%,
4. unloading of the mixture and pressing in hydraulic presses under a pressure of 30 kPa at 60–70°C in order to reduce the fat content in the meal down to 5%,
5. clarification of the fat, i.e. washing with hot water and sedimentation in a clarifier.

In the Stork-Duke equipment the preparation of the raw material and loading to the reactor are automatic. The raw material in the reactor is sterilized, dried, and fragmented at 135–140°C. The continuous reactor is filled only with hot fat (without water). After leaving the reactor the homogeneous mixture is strained through a strainer, and the remaining fat is pressed in a press. The fat is clarified by centrifuging. The system is equipped with an odour remover — the fumes are sucked with a fan through a scrubber and burnt.

1.3.3 Marine fat processing

Marine animal fats, being fish oils and liver oils (the significance of whale fats is gradually decreasing), are by-products obtained during processing of the raw material into food products or feeds, i.e. fish meal. Fish meal production is the main source of marine animal fats. An effluent is obtained during the production of fish meal. This effluent contains among others an oil, which can be directly utilized for technical purposes or can be refined (usually by deproteinization and neutralization). The refined oil obtained is usually hydrogenated. The refining aims therefore mainly at the removal of hydrogenation inhibitors. Post-refining acids are formed during the refining, similarly to the vegetable oil industry.

The yield of fat depends on the fish variety, the production method and the content of the residual fat in the meal. The fatty matter content of the raw fat is usually 95%, and the yield is estimated at 63% on average. Marine animal fat production is significant for world economy, since it oscillates around 1 million ton per year. The main producers are the USA, Peru, Chile, RSA and Japan (Stansby, 1979). There are numerous types of equipment for fish processing, e.g. Centrifish, Atlas, Schloteerhose, Sharples or OHNO. The oil is removed together with the liquid phase in an expeller, a centrifuge, or some other device. In such a system the mixture from the boiling pot passes through an expeller equipped with a drum sieve, letting the effluent out. It is subsequently passed to a supercentrifuge, where it is separated into oil, aqueous phase and suspension residues. Fish oils have high AV and are readily oxidized. Due to this in certain cases it is necessary to thoroughly remove the oxidation products.

Degumming is not always performed as a separate process. The phospholipids are removed together with fatty acids. Sometimes the oil is degummed by means of 2–3% of concentrated tannin solution at a temperature of 35°C, raised at the end to 42–45°C. The FFA content after the neutralization is 0.01–0.03%, while that of phosphorus is 0.5 mg/kg. Removal of colouring substances by alkali neutralization is incomplete and depends on the concentration of the lye. Active carbon in an amount of 0.1–0.3% is used for bleaching. It is often introduced to the hydrogenation autoclave at 150–160°C. The adsorbents remove not only the colouring substances, but also the polymerized matter and traces of metals, which is particularly important in the case of this raw material. The carbon, filtered separately or together with the catalyst, is a useless waste.

1.3.4 Tanning

Fatty wastes from the tanning industry comprise cow hide wastes, so-called shavings, obtained after soaking, and pig skin wastes — grease from soaking of skins, wastes from shaving of soaked skin and limed skin, as well as sewage greases. Fatty raw material from tanning has various properties and stability depending on the technological stage at which it is obtained. The least stable is the raw material from soaking. Due to its low salinity it is readily attacked by microflora and undergoes decay. For this reason it should be rendered as soon as possible.

1.3.5 Bone processing

Bone fat is obtained as a by-product during processing of raw or stored bones into gelatin, bone meals and glues. It can be utilized both as an alimentary and as a technical raw material. The quality of bone fat depends on the kind and freshness of the raw material, as well as the method of bone processing.

Figure 1.56 presents a schematic diagram of the production of edible and non-edible bone fats.

It is accepted that the fat content in bones is 13% on average, and that the yield is equal to 94.5% of edible fat or 92.3% of non-edible fat. The fat content in the raw edible fat is ca. 98%, while in the non-edible fat — 95%.

Non-edible fat is obtained from degreased waste bones mainly by solvent extraction methods or by a wet method.

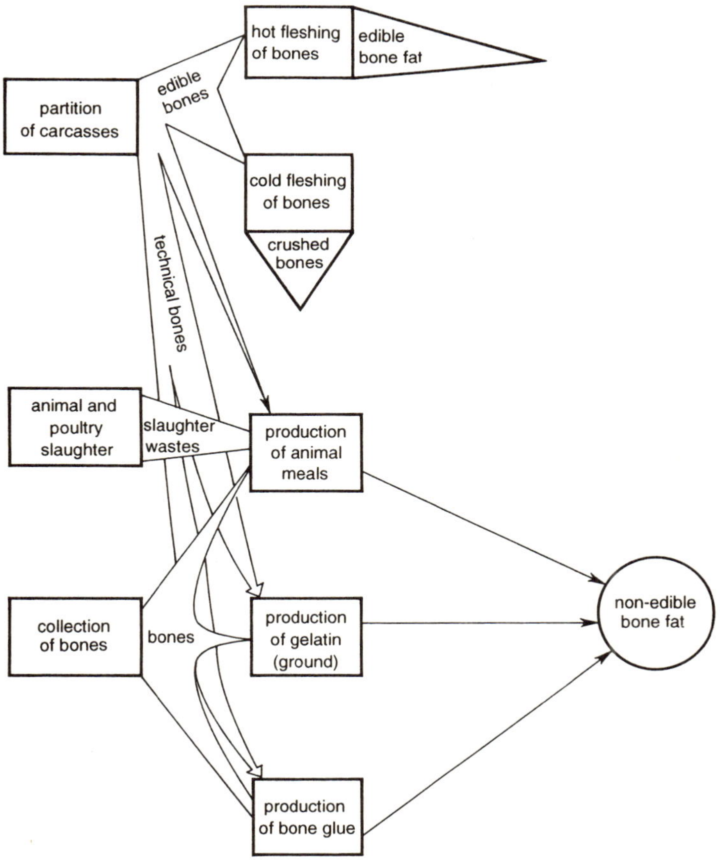

Fig. 1.56 Outline of bone fat obtaining (Urban, 1979).

1.3.6 Poultry processing

Poultry fat is obtained by water or solvent extraction from all the wastes formed during industrial poultry processing and is a by-product from this processing. The composition and the characteristics of this material depend on the method used to obtain it. Aqueous extraction yields brighter oil, but of high oxidation degree (Szczepa°ska *et al.*, 1979). Due to a large content of impurities — particularly proteinaceous — this raw material requires refining prior to further processing.

1.3.7 Others

1.3.7.1 Breeding of furred animals

Breeding of furred animals, like foxes, minks, coypu, etc., can be a source of good quality rendered animal fats. There are two methods of utilization of the waste after skinning of the furred animals. The first consists in processing the whole carcasses into utilization grease and feed meals, while the second involves removal of the subcutaneous adipose tissue and the adipose tissue from the carcasses directly after skinning, followed by subsequent rendering of the fat (Kosko, 1983). It has been established that an average of 1.4 kg of fat can be obtained from a polar fox, 0.9 kg from a common fox, and 0.25 kg from a mink.

Dry rendering of the adipose tissue at a temperature not exceeding 80°C produces fat with a yield of 78–85% with respect to the tissue, and 9–11% of cracklings.

1.3.7.2 Sewer grease

Sewer greases are obtained from rendering of raw sewer greases from rendering plants, meat processing plants, utilization plants and mass nutrition facilities, as well as from municipal wastes. They can be a typical example of wastes that pollute the natural environment when not utilized.

Liquid wastes from slaughterhouses are one of the basic sources of sewer greases. According to Rutkowski (1974), 7–9 m^3 of liquid wastes are produced per 1 t of slaughter animals in the USSR; 15 m^3 of water is used per 1 t of slaughter animals in a slaughterhouse in Chicago, 4 m^3 per animal in Germany, while in the United States — 1.5 m^3 per head of cattle and 0.54 m^3 per swine. Large amounts of fats are disposed of together with the wastes. It is estimated that the amount of fat in the liquid wastes is equal to 1.2–1.8 kg per head of cattle and to 0.4–0.6 kg per swine. The presence of fat in the sewage system causes blockage of the devices and hinders the biological processes of waste treatment.

Skimming tanks are used to recover the majority of the fatty matter from the wastes. They utilize the differences in the density of water and fat. The process is followed by further purification, e.g. by flotation or some other method. The average yield of fatty matter from raw sewer grease is ca. 4%.

The frequency of emptying the tank significantly influences the fatty matter content of raw sewer grease — longer skimming results in a reduced amount of water, yet the tank should not be overfilled, since this decreases the clarity of the effluent. Too long skimming adversely affects the quality of the fat (Table 1.32).

Table 1.32 Changes in properties of fatty matter in crude sewer grease during skimming in a skimming tank (Rutkowski, 1974)

Kind of sample	IV	SV	AV	m.p. (°C)
Grease reaching the skimming tank	48	192	3.4	32.7
Grease after skimming for:				
1 day	48	192	4.5	32.7
14 days	42	192	128	36.3
21 days	34	192	167	38.7
45 days	21	189	169.9	46.3

The following methods are used to obtain the non-edible fat from this kind of a raw material:

— rendering coupled with saponification of the glycerides with caustic soda, followed by liberation of fatty acids by acidification,
— typical thermal rendering of fats,
— heating in the presence of sulphuric acid (Zamyščajaeva *et al.*, 1974).

Saponification requires large amounts of chemicals and produces much waste, containing large amounts of sodium sulphate. Fatty acids of colour equal to 110 in the iodine scale, a content of ether solubles equal to 95.1%, AV equal to 183.1 and SV — to 180.4, were obtained from crude sewer grease containing 86.7% of ether solubles and 1.3% of UM, of colour equal to 1076 in the iodine scale and SV equal to 176.1. The crude grease was saponified at 80°C with NaOH (density 1.222) and acidified with 10% H_2SO_4 (Rutkowski, 1974).

Heating of crude sewer grease results in relatively easy breaking of emulsions and separation of the fatty phase. After straining the coarse impurities on a sieve, the fat is rendered without or in the presence of brine and phase separation occurs. Brine is used to increase the density difference between the aqueous and fatty phases and to increase the solubility of a part of protein substances in the salt solution, which accelerates the separation (Table 1.32). Proper separation of fat from strongly contaminated water is the most important stage of the process. Centrifuging is the most efficient in cases of strong contamination, provided the centrifuge is well protected against damage by the coarse impurities present in the wastes. Water containing many organic impurities is a waste recycled through a fat trap.

In cases when the crude sewer grease originates from sewage treatment by coagulation, it is recommended to heat it and to add sulphuric acid to it. Fatty matter isolated from the raw material whose characteristics was given in Table 1.33 contained 4.8% UM, 23.3% FA and 52% oxidized FA. The aqueous phase contained regenerated aluminium sulphate, which could have been recycled to coagulation.

Table 1.33 Characteristics of sewer grease rendered in and without the presence of brine (Rutkowski, 1974)

Rendering conditions fat : water ratio	Components and properties of grease				
	Fatty matter (%)	Water (%)	Ash (%)	AV	Colour in iodine scale
with 1% brine					
1:2	98.8	0.6	0.06	23.4	58
1:1	98.9	0.5	0.08	23.7	45
1:0.5	98.8	0.6	0.09	23.7	58
without brine					
1:2	98.3	0.7	0.05	23.4	58
1:1	98.2	0.7	0.06	23.5	58
1:0.5	98.1	0.8	0.07	23.6	45

The composition of crude sewer grease obtained from liquid wastes purified by flotation with aluminium sulphate as a coagulating agent was the following (Zamyščajaeva *et al.*, 1974):

Components	Content (%)	FA composition	Weight %
water and volatile substances	86–87	C_{12}	0.6–2.5
dry matter	12.8–13.8	C_{14}	1.3–25.5
inorganic compounds	0.6–2.5	C_{16}	traces–23.5
organic compounds	10.3–13.2	$C_{18:1}$	31–32.5
UM	0.4–1.7	$C_{18:2\ \text{isolated}}$	15.6–25.6
FA	8.1–9.4	$C_{18:2\ \text{conjugated}}$	0.4–1.9
oxidized FA	0.7–1.5		

REFERENCES

Abrachimov A. *et al.* (1981), *Masložr. Prom.*, **4**, 17.

Annesini M. C., Wova A. R., Gironi F. and Pochetti F. (1983), *Effluents & Water Treatment*, **6**, 245–248.

Anon (1973), *Soap, Perf. and Cosmetics*, **9**, 527.

Athanassiadis A. (1982), *J. Am. Oil Chem. Soc.*, **59**, 554.

Baranowski J. and Busko-Baranowska A. (1981), *Polish Patent Appl.* No. P-233939.

Basu P.K. (1976), *Chem. Age of India*, **27**, 871.

Beach D. (1983), "Rapeseed crushing and extraction", in: *High and Low Erucic Rapeseed Oils*, Kramer J. *et al.* (Eds.) Academic Press, Toronto.

Beal R. (1981), *Proc. Second A.S.A. Symposium on Soyabean Processing*, Am. Soyabean Ass., Antwerp.

Becker K. (1983), *J. Am. Oil Chem. Soc.*, **60**, 216.

Berger K. G. (1983), *J. Am. Oil Chem. Soc.*, **60**, 158–162.

Berger R. and McPerson W. (1979), *J. Am. Oil Chem. Soc.*, **56**, 743A.

Bernardini R. (1973), *The New Oil and Fat Technology*, Publ. House Techn., Roma.

Bhownick D. N. and Sarma S. A. N. (1977), *Ind. Eng. Chem.*, **16**, 107.

Billenstein S. and Blaschke G. (1984), *J. Am. Oil Chem. Soc.*, **61**, 353.

Braae B. (1976), *J. Am. Oil Chem. Soc.*, **53**, 353.

Brekke O. (1980a), "Oil degumming and soyabean lecithin", in: *Handbook of Soy Oil Processing* and Utilization, Erickson D. *et al.* (Eds.), Am. Soyabean Ass., St. Louis, and Am. Oil Chem. Soc., Champaign, 71.

Brekke O. (1980b), "Deodorization", in: *Handbook of Soy Oil Processing and Utilization*, Erickson D. *et al.* (Eds.), Am. Soyabean Ass., St. Louis and Am. Oil Chem. Soc., Champaign, 155.

Brian K. (1976), *J. Am. Oil Chem. Soc.*, **53**, 27.

Busujuščij V. N. *et al.* (1970), *Masložir. Prom.*, **10**, 19.

Bytenev V. P. *et al.* (1980), *Masložir. Prom.*, **12**, 22.

Cahn A. (1979), *J. Am. Oil Chem. Soc.*, **56**, 809A.

Cavanagh G (1976), *J. Am. Oil Chem. Soc.*, **53**, 361.

Cavanagh G. (1988), *J. Am. Oil Chem. Soc.*, **65**, 512.

Cechnicki J. and Kosmicki W. (1970), *Pollena*, **14**, 132.

Cegłowska K. and Terlewicz H. (1979), *X Sesja Naukowa Kom. Technol. Chem. i Żywn.*, PAN, Kraków, abstracts, p. 265.

Chaudry M. *et al.* (1976), *J. Am. Oil Chem. Soc.*, **53**, 695.

Chone E. (1985), *Bulletin Cetiom*, No. 90, 1st trimestre.

Crauer L. (1970), *J. Am. Oil Chem. Soc.*, **47**, 210A.

Černyševa D. A. and Šmidt A. A. (1975), *Masložir. Prom.*, **3**, 17.

Daun K. (1982), "Oilseeds-processing", in: *Grains and Oilseeds*, Canadian International Grains Institute, Winnipeg.

Davidsohn A. S. and Moreth G. F. (1985), *Seifen, Öle, Fette, Wachse*, **111**, 265.

Dijkstra A. J. and Van Opstal M. (1989), *Proceedings of World Conference on Edible Fats and Oils Processing*, pp. 176–178, Ed. Erickson D. R., American Oil Chemists Society.

Diosady L. *et al.* (1984), *J. Am. Oil Chem. Soc.*, **61**, 1366.

Dupjohann J. and Hemford H. (1975), *Fette, Seifen, Anstrichm.*, **77**, 81.

Efimov *et al.* (1980), *Masložir. Prom.*, **7**, 24.

El-Shami S. M. *et al.* (1985), *Seifen, Öle, Fette, Wachse*, **111**, 415.

Fajviševskij M. L. and Mginaridze T. D. (1976), *Masložir. Prom.*, **2**, 20.

Farris R.D. (1979), *J. Am. Oil Chem. Soc.*, **56**, 770A.

Fattori *et al.* (1988), *J. Am. Oil Chem. Soc.*, **65**, 968.

Fetzer W. (1983), *J. Am. Oil Chem. Soc.*, **60**, 203.

Fischer N. (1981), *Fette, Seifen, Anstrichm.*, **83**, 507.

Florin S. and Bartesch H. (1983), *J. Am. Oil Chem. Soc.*, **60**, 193.

Forster A. and Harper A. (1983), *J. Am. Oil Chem. Soc.*, **60**, 265.

Friedrich J. and Pryde E. (1984), *J. Am. Oil Chem. Soc.*, **61**, 223.

Friedrich J. P., List G. R. and Heakin A. J. (1982), *J. Am. Oil Chem. Soc.*, **61**, 1849–1851.

Fullbrook P. (1983), *J. Am. Oil Chem. Soc.*, **60**, 476.

Gavin A. (1978), *J. Am. Oil Chem. Soc.*, **55**, 783.

Gavrilko N. P. *et al.* (1976), *Masložir. Prom.*, **2**, 23.

Gorbačeva R. A. *et al.* (1978a), *Masložir. Prom.*, **5**, 17.

Gorbačeva R. A. *et al.* (1978b), *Masložir. Prom.*, **9**, 18.

Grecki W. (1961), *Tłuszcze, Środki Piorące*, **1**, 50.

Grothues B. (1982), *Fette, Seifen, Anstrichm.*, **84**, 21.

Grundmann J. (1984), *Fette, Seifen, Anstrichm.*, **86**, 554.

Haraldsson G. (1983), *J. Am. Oil Chem. Soc.*, **60**, 251.
Haraldsson G. (1984), *J. Am. Oil Chem. Soc.*, **61**, 219.
Heilman J. (1983), *J. Am. Oil Chem. Soc.*, **60**, 248.
Hinze A. G. (1984), *Fette, Seifen, Anstrichm.*, **86**, 520.
Holmbom B. (1978a), *J. Am. Oil Chem. Soc.*, **55**, 876.
Holmbom B. (1978b), *Constitutents of Tall Oil*, Acad. dissertation, Inst. of Wood Chem. and Cellulose Techn., Finland.
Homan T. *et al.* (1981), *Fette, Seifen, Anstrichm.*, **83**, 570.
Hoq M. M. *et al.* (1984), *J. Am. Oil Chem. Soc.*, **61**, 776.
Hornus P. and Guimjeu E. N. (1992), *Oleagineux*, **47**, 250–254.
Ilsemann K. and Mukherjee K. D. (1978), *J. Am. Chem. Soc.*, **55**, 892.
Irodov M. V. (1970), *Masložir. Prom.*, **9**, 14.
Irodov M. V. (1973), *Masložir. Prom.*, **1**, 22.
Irodov M. V. *et al.* (1971), *Masložir. Prom.*, **7**, 73.
Irodov M. V. *et al.* (1976), *Masložir. Prom.*, **1**, 21.
Johnson R. W. (1984), *J. Am. Oil Chem. Soc.*, **61**, 241.
Johnson L. and Lusas E. (1983), *J. Am. Oil Chem. Soc.*, **60**, 229.
Kadesh R. G. (1979), *J. Am. Oil Chem. Soc.*, **56**, 845A.
Kalam A. and Joshi J. B. (1988), *J. Am. Oil Chem. Soc.*, **66**, 1536–1540.
Kamyšan M. A. and Derevjanko J. A. (1972), *Masložir. Prom.*, **5**, 37.
Kehse W. (1976), *Fette, Seifen, Anstrichm.*, **78**, 50.
Kelly M. J. (1980), *J. Am. Oil Chem. Soc.*, **57**, 116.
Klimmek H. (1984), *J. Am. Oil Chem. Soc.*, **61**, 200.
Kločko N. D. and Dorofeeva N. S. (1972), *Masložir. Prom.*, **5**, 17.
Kločko N. D. and Vetkazova S. A. (1976), *Masložir. Prom.*, **6**, 25.
Knoer P. (1970), *Fette, Seifen, Anstrichm.*, **72**, 1066.
Kock M. (1983), *J. Am. Oil Chem. Soc.*, **60**, 198.
Kocór M. and Kroszczyński (1978), unpublished work.
Kopylenko S. D. *et al.* (1979), *Masložir. Prom.*, **4**, 39.
Kosko I. (1983), *Zesz. Probl. Post. Nauk Rol.*, **32**, 147.
Kozłowska H., Nowak H. and Zadernowski R. (1988), *Fette, Seifen, Anstrichm.*, **90**, 216.
Krause A. (1978), *Fette, Seifen, Anstrichm.*, **80**, 77.
Kreutzer U. R. (1984), *J. Am. Oil Chem. Soc.*, **61**, 343.
Kroll S. (1980), *Fette, Seifen, Anstrichm.*, **82**, 357.
Kubicki M. (1983), *Tłuszcze Jadalne*, **21** (4), 28.
Kubicki M. and Wojnarowicz C. (1973), *Tłuszcze Jadalne*, **17**, 88.
Lanteri A. (1978), *J. Am. Oil Chem. Soc.*, **55**, 128.
Latondress E. (1983), *J. Am. Oil Chem. Soc.*, **60**, 257.
Leonard E. C. (1979), *J. Am. Oil Chem. Soc.*, **56**, 782A.
Leonard E. C. and Kopald S. L. (1984), *J. Am. Oil Chem. Soc.*, **61**, 176.
Leščenko P. S. and Levit M. S. (1973), *Masložir. Prom.*, **5**, 28.
Levit M. S. *et al.* (1971), *Masložir. Prom.*, **2**, 36.
Liebing H. and Lau J. (1976), *Fette, Seifen, Anstrichm.*, **78**, 123.
Liebing H. *et al.* (1983), *Fette, Seifen, Anstrichm.*, **85**, 289.
Lim C. L. (1988), *Proceedings of International Oil Palm Conference*, Kuala Lumpur, June 1987, pp. 155–176, Palm Oil Research Institute of Malaysia.
Linfield W. M. *et al.* (1984), *J. Am. Oil Chem. Soc.*, **61**, 1067.
List G. R., Friedrich J. P. and Christianson D. D. (1984a), *J. Am. Oil Chem. Soc.*, **61**, 1849–1851.
List G. R., Friedrich J. P. and Pominski J. (1984b), *J. Am. Oil Chem. Soc.*, **61**, 1847–1849.
Logan R. L. (1979), *J. Am. Oil Chem. Soc.*, **56**, 779A.
Lower E. S. and Ashworth D.R. (1974), German Patent Appl. 2353274.
Łuczyn S. *et al.* (1981), Polish Pat. 109468.
Maclellan M. (1983), *J. Am. Oil Chem. Soc.*, **60**, 368.
Maerker G. (1979), *J. Am. Oil Chem. Soc.*, **56**, 726A.
Mag T. (1983), *J. Am. Oil Chem. Soc.*, **60**, 380.
Mag T. *et al.* (1983), *J. Am. Oil Chem. Soc.*, **60**, 1008.

Mangold H. (1983), *J. Am. Oil Chem. Soc.*, **60**, 226.
Masłowska J. and Dorabialski A. (1979), *X Sesja Naukowa Kom. Technol. Chem. Żywn.*, PAN, Kraków, abstracts, p. 313.
Masłowska J. and Baranowski J. (1979), *X Sesja Naukowa Kom. Technol. Chem. Żywn.*, PAN, Kraków, abstracts, p. 308.
Mazgajska I. and Smenda J. (1978), unpublished work.
Mazgajska I. *et al.* (1978), *Chemik*, **10**, 315.
Mazgajska I. *et al.* (1979), *X Sesja Naukowa Kom. Technol. Chem. Żywn*, PAN, Kraków, abstracts, p. 318.
Meffert A. (1984), *J. Am. Oil Chem. Soc.*, **61**, 265.
Melnikov K. A. and Mnuchin U. J. (1980), *Masložir. Prom.*, **12**, 26.
Molčanov I. V. and Evdokimov G. M. (1971), *Masložir. Prom.*, **8**, 18.
Monick J. A. (1979), *J. Am. Oil Chem. Soc.*, **56**, 853A.
Moore H. (1983), *J. Am. Oil Chem. Soc.*, **60**, 189.
Moore H. and Yeates J. (1979), US Patent 4,154,759.
Morren J. (1981), US Patent 3,428,660.
Morrison W. (1982), *J. Am. Oil Chem. Soc.*, **59**, 519.
Mounts T. (1983), *J. Am. Oil Chem. Soc.*, **60**, 453.
Myers N. (1983), *J. Am. Oil Chem. Soc.*, **6**, 224.
Nabiev A. C. and Isaev Ch. J. (1978), *Masložir. Prom.*, **10**, 27.
Niedbalska W. and Witwicka J. (1977), unpublished work.
Nieuwenhuyzen W. (1976), *J. Am. Oil Chem. Soc.*, **53**, 425.
Niewiadomski H. (1979), *Edible Fat Technology* (in Polish), WNT, Warszawa.
Niewiadomski H. (1983), *Rapeseed Technology* (in Polish), PWN, Warszawa.
Nitsche M. (1984), *Fette, Seifen, Anstrichm.*, **86**, 117.
Ogilec M. W. and Ložešnik W. K. (1981), USSR Patent 806751.
Ong J. (1983), *J. Am. Oil Chem. Soc.*, **60**, 314.
Ong J. and Sinkeldam E. (1983), *Fette, Seifen, Anstrichm.*, **85**, 304.
Pardun H. (1979), *Fette, Seifen, Anstrichm.*, **81**, 297.
Pardun H. (1983), *Fette, Seifen, Anstrichm.*, **85**, 1.
Patterson H. (1976), *J. Am. Oil Chem. Soc.*, **53**, 339.
Penk G. (1981a), *Fette, Seifen, Anstrichm.*, **83**, 558.
Penk G. (1981b), *Proc. Second A.S.A. Symposium on Soyabean Processing*, Antwerp
Poley E. (1957), US Patent 2,784,161.
Porybleva T. P. and Zajceva L. V. (1976), *Masložir. Prom.*, **11**, 19.
Posorske L. H. (1984), *J. Am. Oil Chem. Soc.*, **61**, 1758.
Prokopenko L. G. (1978), *Masložir. Prom.*, **8**, 39.
Reck R. A. (1979), *J. Am. Oil Chem. Soc.*, **56**, 796A.
Red J. and Ilagen J. (1978), US Patent 4,118,407.
Rojzmann B. B. *et al.* (1971), *Masložir. Prom.*, **5**, 21.
Richtler H. J. and Knaut J. (1984), *J. Am. Oil Chem. Soc.*, **61**, 160.
Ringers H. and Segers J., US Patent 4,049,686. D.A.S. 2609705.
Roden A. (1983), "Some recent innovation in Canola processing technology", in: *High and Low Erucic Acid Rapeseed Oils*, Kramer J. *et al.* (Eds.), Academic Press, Toronto, p. 553.
Ruckenstein C. (1982), *Fette, Seifen, Anstrichm.*, **84**, 504.
Rutkowski A. (1974), unpublished work.
Rutkowski A. and Szczepańska H. (1971), *Seifen, Öle, Fette, Wachse*, **97**, 716.
Rutkowski A. *et al.* (1978), *Riv. Ital. Sost. Grasses*, **15**, 389.
Sandving H. (1983), *J. Am. Oil Chem. Soc.*, **60**, 243.
Schumacher H. (1983), *Fette, Seifen, Anstrichm.*, **85**, 220.
Segers J. (1982), *Fette, Seifen, Anstrichm.*, **84**, 543.
Segers J. (1983), *J. Am. Oil Chem. Soc.*, **60**, 262.
Segers J. C. and Van de Sande R. L. K. M. (1990), *Proceedings of World Conference on Edible Fats and Oils Processing*, Maastricht, October 1989, pp. 99–93, Ed. D. R. Erickson, American Oil Chemists Society.
Senda H. *et al.* (1975), Japan Patent 7537802.

Signoret A. and Vermeersch G. (1988), *Rev. Franc. Corps Gras*, **35**, 391–396.

Sikorski H. and Staniewski J. (1976), *Pollena*, **20**, 241.

Smirnov A. P. (1972), *Masložir. Prom.*, **7**, 40.

Sonntag N. O. V. (1979a), *J. Am. Oil Chem. Soc.*, **56**, 729A.

Sonntag N. O. V. (1979b), *J. Am. Oil Chem. Soc.*, **56**, 861A.

Sonntag N. O. V. (1981), *J. Am. Oil Chem. Soc.*, **58**, 155A.

Sonntag N. O. V. (1982), *J. Am. Oil Chem. Soc.*, **59**, 795A.

Sonntag N. O. V. (1984), *J. Am. Oil Chem. Soc.*, **61**, 229.

Stage H. (1973), *Fette, Seifen, Anstrichm.*, **75**, 160.

Stage H. (1975), *Riv. Ital. Sost. Grasses*, **52**, 291.

Stage H. (1978a), *Fette, Seifen, Anstrichm.*, **80**, 257.

Stage H. (1978b), *Seifen, Öle, Fette, Wachse*, **104**, 245.

Stage H. (1982), *Fette, Seifen, Anstrichm.*, **84**, 333.

Stage H. (1984a), *J. Am. Oil Chem. Soc.*, **61**, 204.

Stage H. (1984b), *Fette, Seifen, Anstrichm.*, **86**, 541.

Stage H. and Bose K. (1974), *Chem. Age of India*, **25**, 581.

Stansby M. E. (1979), *J. Am. Oil Chem. Soc.*, **56**, 793A.

Stockburger G. J. (1979), *J. Am. Oil Chem. Soc.*, **56**, 774A.

Surewicz W. and Surma-Ślusarska B. (1983), unpublished work.

Swern D. (1964), *Bailey's Industrial Oil and Fat Products*, Interscience Publ., New York.

Szemraj H. (1973), *Tłuszcze Jadalne*, **17**, 169.

Szczepańska H. (1983), *Zesz. Prob. Post. Nauk Rol.*, **274**, 73.

Szczepańska H. and Szelejewski W. (1979), unpublished work.

Szczepańska H. *et al.* (1971), unpublished work.

Szczepańska H. *et al.* (1972), *Przemysł Chem.*, **51**, 211.

Szczepańska H. *et al.* (1979), *Pollena*, **23**, 123.

Szczepańska H. *et al.* (1982), *Przemysł Chem.*, **61**, 245.

Szuhaj B. (1983), *J. Am. Oil Chem. Soc.*, **60**, 306.

Tandy O. and McPherson W. (1980), in: *Handbook of Soy Oil Processing and Utilization*, Erickson D. *et al.* (Eds.), Am. Soyabean Ass., St. Louis and Am. Oil Chem. Soc., Champaign, p. 217.

Teasdale B. and Mag T. (1983), "The commercial processing of low and high erucic acid rapeseed oil", in: *High and Low Erucic Acid Rapeseed Oils*, Kramer J. *et al.* (Eds.), Academic Press, Toronto.

Teramoto T. *et al.* (1973), Japan Patent 7303354.

Tranchino L., Melle F. and Sodini G. (1984), *J. Am. Oil Chem. Soc.*, **61**, 1261–1263.

Trosko U. I. *et al.* (1977), *Masložir. Prom.*, **7**, 27.

Urban R. (1979), *Outline and principles of a complex balance of fats in Poland. A study of* the *team of experts of Chief Technical Organization No. 5454* (in Polish), Warszawa.

Van Haften J. L. (1979), *J. Am. Oil Chem. Soc.*, **56**, 831A.

Vasileva L. G. *et al.* (1974), *Masložir. Prom.*, **9**, 27.

Volotovskaya S. N. and Grin G. (1982), *Masložir. Prom.*, **3**, 20.

Volotovskaya S. N. *et al.* (1973), *Masložir. Prom.*, **10**, 13.

Volotovskaya S. N. *et al.* (1979), *Masložir. Prom.*, **2**, 16.

Volotovskaya S. N. *et al.* (1982), USSR Patent 897841.

Waldmann C. and Eggers R. (1991), *J. Am. Oil Chem. Soc.*, **68**, 922–930.

Walisiewicz-Niedbalska W. and Witwicka J. (1977), unpublished work.

Walisiewicz-Niedbalska W. and Witwicka J., in print.

Ward J. (1984), *J. Am. Oil Chem. Soc.*, **61**, 1358.

Welsh W. A., Bogdanor J. M. and Toeneboehn G. J. (1990), *Proceedings of World Conference on Edible Fats and Oils Processing*, 1989, pp. 189–202, Ed. D. R. Erickson, American Oil Chemists Society.

Weber K. (1981), *Fette, Seifen, Anstrichm.*, **85**, 544.

Wendt H. (1981), *Fette, Seifen, Anstrichm.*, **83**, 541.

Witwicka J. *et al.* (1979), *X Sesja Nauk. Kom. Technol. Chem. Żywn.*, PAN, Kraków, abstracts, p. 359.

Woerfel J. (1983), *J. Am. Oil Chem. Soc.*, **60**, 310.

Wolff J. (1983), *J. Am. Oil Chem. Soc.*, **60**, 220.
Wong M. (1983), *J. Am. Oil Chem. Soc.*, **60**, 316.
Young F. (1983), *J. Am. Oil Chem. Soc.*, **60**, 374.
Zamyščajaeva M. *et al.* (1974), *Maslož̌ir. Prom.*, **7**, 40.
Zockoll C. (1982), *Fette, Seifen, Anstrichm.*, **84**, 558.
Zschau W. (1979), *Fette, Seifen, Anstrichm.*, **81**, 303.

CHAPTER 2

Composition, Properties and Utilization of By-Products

2.1 OIL INDUSTRY

By-products and wastes from the oil industry play an important role as raw materials. The most important is the solvent extracted meal, used as a feed, while fatty products, like post-refining acids, oil extracted from oiled bleaching earth and deodorizer scum, are used for technical purposes.

2.1.1 Meal

Three of five main oily raw materials, viz, soyabean, rapeseed and cotton seed, contain anti-nutritive components. An increasing demand for nutritive fodders results in a widespread utilization of not only sunflower and peanut meals, but also meals obtained from the above-mentioned three plants as sources of protein. However, in the latter case either new varieties free from the unwanted components are bred, or it is necessary to process the meal in a special way or to reduce the food rations. Utilization of peanut and cotton-seed meals creates also another danger — the presence of aflatoxins produced by *Aspergillus flavus* mould. According to the proposed EEC regulations the content of aflatoxins cannot exceed 0.02 mg/kg in food additives for milk cows. In terms of the average aflatoxin content in cotton-seed meal this means a maximum addition of 13% to the mash, while in the case of peanut meal — up to 4%. The permissible content of free gossypol in cotton-seed meal is 500 mg/kg for milk cows and 100 mg/kg for poultry. In terms of the percentage of this meal in fodders these yield 3.3% and 0.7%, respectively. The above limits make the profitability of using these meals as feeds doubtful. In the light of these data application of any technological operations and processes aiming at the removal of the unwanted components from the meals should be investigated taking into account the economical aspects (Roberts, 1976).

After leaving the extractor the meal is broken down to particles of ca. 2 mm size. Sometimes it is dusty, unpalatable and difficult to handle.

To overcome these disadvantages it is pelletized by applying elevated temperature, humidity and pressure. The pellets are of ca. 6 mm size. They can be packed more easily and are more agreeable for consumption. After forming it is necessary to condition them in order to remove the excessive humidity and to cool them down. Conditioning is carried out in a horizontal or vertical cooler. The humidity should be 12%, while the temperature should be higher by 8°C than ambient.

Meal is used for some purposes in a granulated form. The granulated product is obtained most often directly after cooling the pellets. Meal in this form is used as a starter fodder for chicken and laying hens. After breaking, the material is screened through sieves of a suitable mesh size. The most proper way of supplementing the meal with fat is the direct introduction of the fat to pellets or to the granulated product.

Due to high labour costs the meal is increasingly often sold in bulk.

Rapeseed meal is widely used in Europe and Canada as a high-protein fodder component. However, its utilization is limited due to the presence of glucosinolates in the traditional varieties. The glucosinolates are hydrolyzed by an enzyme myrosinase, forming thiocyanates, nitriles and vinyloxazolidinethione (VTO). These compounds induce goitre in animals fed with excessive doses of this meal. This, in turn, hinders the synthesis of thyroxine. The formed VTO influences the thyroid hormones secretion. To remove these detrimental components the seed or the meal is subject to various technological processes. These processes, however, adversely influence proteins, thus decreasing the nutritional value of the meal. To overcome this problem a rapeseed variety of a reduced glucosinolate content has been bred.

Cellulose is another component reducing the usefulness of meal. Its main occurrence is in hull, which is 1/3 cellulose. Hull contains also polymerized complexes of phenols and proteins. These complexes are insoluble and have no nutritive value for non-ruminants and poultry. Therefore it is recommended to remove hull. Meal from decorticated double zero rapeseed can compete with soyabean meal both with respect to the protein content, and to the nutritive value. This is due to the fact that no vegetable protein has an ideal composition of the essential amino acids, particularly those containing sulphur and lysine, which in cereal or leguminous proteins have a less than optimal share. For this reason rapeseed protein is interesting also for human nutrition. It can be utilized for the production of flour, concentrates and isolates. The following is an example of the composition of a concentrate from Tower variety protein (Daun *et al.*, 1982):

	%	mg/kg
fat	0.9	ITC < 0.02
protein	66.2	VTO< 0.06
cellulose	10.5	
ash	6.6	
nitrogen-free extract	15.8	

Table 2.1 Basic composition of selected meals (Daun *et al.*, 1982)

Meal	Content (%)						
	crude protein	fat	cellulose	Ca	P	ash	dry matter
Cotton-seed *	41.0	3.9	12.6	0.17	0.97	6.2	91.4
Cotton-seed **	41.0	2.1	11.3	0.16	1.0	6.4	90.4
Cotton-seed ***	41.0	0.81	12.7	0.17	1.0	6.4	89.9
Linseed *	32.0	3.5	9.5	0.4	0.8	6.0	90.0
Linseed **	33.0	0.5	9.5	0.35	0.75	6.0	88.3
Peanut **	47.0	1.2	13.1	0.2	0.65	4.5	91.5
Rapeseed *	36.0	6.7	12.2	0.71	1.0	6.8	94.0
Rapeseed **	36.0	2.6	13.2	0.66	0.93	7.2	92.0
Rapeseed ***	38.0	3.8	11.1	0.68	1.17	7.2	91.2
Sunflower *	41.0	7.6	13.0	0.43	1.0	6.8	93.0
Sunflower **	42.0	2.3	13.0	0.4	1.0	7.7	93.0
Soyabean *	42.0	3.5	6.5	0.2	0.6	6.0	89.0
Soyabean *	44.0	0.5	7.0	0.25	0.6	6.0	89.6
Herring	72.0	10.0	1.0	2.0	1.0	10.4	93.0
Menhaden	62.0	10.2	1.0	5.0	3.0	20.0	92.0

* From pressing.
** From solvent extraction.
*** From pressing and solvent extraction.

Table 2.2 A comparison of the content of some more important amino acids in meal from solvent extraction of various raw materials (Daun *et al.*, 1982)

Amino acids	Content (%) in d.m. of meal		
	rapeseed	soyabean	sunflower
Methionine	0.7	0.65	1.5
Cystine	0.47	0.67	0.7
Lysine	2.3	2.9	1.7
Tryptophane	0.44	0.7	0.5
Threonine	1.71	1.7	1.5
Isoleucine	1.51	2.5	2.1
Histidine	1.2	1.1	1.0
Valine	1.94	2.4	2.3
Leucine	2.6	3.4	2.6
Arginine	2.3	3.4	3.5
Phenyl alanine	1.5	2.2	2.2
Glycine	1.9	2.4	2.7

The chemical composition of the meal depends on the composition of the raw materials and to some extent on the degree of oil removal. Table 2.1 lists the principal composition of meals used most often for feeding purposes, while Tables 2.2, 2.3 and 2.4 give the content of amino acids, vitamins and mineral components of rapeseed, soyabean and sunflower meals. On the basis of these data the share of the particular meals in mash can be determined. Metals are significant components of meals. It follows

Table 2.3 A comparison of the vitamin content in meal from solvent extraction of various raw materials (Daun *et al.*, 1982)

Vitamins	Content in meal (mg/kg)		
	rapeseed	soyabean	sunflower
E	—	3.0	11.0
Thiamine	5.2	1.7	—
Riboflavine	3.7	3.0	3.1
Pantothenic acid	9.5	13.3	10.0
Biotin	0.9	0.32	—
Folic acid	2.3	0.45	—
Choline	6700	2743	2900
Niacin	159.5	59.8	220.0

Table 2.4 A comparison of the content of mineral components in meal from solvent extraction of various raw materials (Bragg, 1974; Daun *et al.*, 1982)

Components	Content in meal		
	00 rapeseed	soyabean	sunflower
		%	
Sodium		0.027	2.0
Potassium	1.29	1.97	1.0
Magnesium	0.64	0.27	0.75
Sulphur		0.43	
Calcium	0.61	0.32	
Phosphorus	1.88	0.67	
		mg/kg	
Manganese	54	27.5	23.0
Iron	159	120	34.2
Copper	10.4	36.3	3.5
Zinc	71.4	27.0	
Selenium	1.0	0.1	

from Table 2.4 that rapeseed meal is richer in Ca, P, Zn, Fe, Mn, and Se than soyabean meal.

The usefulness of rapeseed meal as a fodder depends to some extent on the manner of extraction of the fat from the seed (Bell and Jeffers, 1976). Following is a comparison of favourable and unfavourable properties of soyabean and rapeseed meals (Hussar, 1977):

Maximal contents of rapeseed meal in feeds are given by Clendinin *et al.* (1978, 1981).

Rapeseed meal of high glucosinolate content was enriched with fat by introduction of post-refining fatty acids, oil sludges, oil from skimming

Nutritive agents	Favourable properties of the meal	
	soyabean	rapeseed
Crude protein	30% more	
Amino acids (per unit protein)		
lysine	10% more	
methionine		20% more
Calcium		100% more
Phosphorus		ca.100% more
Vitamins B		more niacin and choline
Crude cellulose	ca. 50% less	
Energetic value	10–20% higher	
Anti-nutritive agents	**Manner of eliminating the unfavourable properties**	
Urease, proteinases	thermal inactivation	
Goitrins		thermal inactivation of myrosinase liberating goitrin

tanks and oiled bleaching earth. Fat content in such meal was 5.6–10.3%. Investigations on the effect of storage of such meals under industrial conditions revealed that within one month they do not undergo unfavourable changes, either chemical and sensory (Minkowski *et al.*, 1978).

The usefulness of fractionation of well ground rapeseed meal into flour and hulls has been examined on a semi-commercial scale (Sosulski and Zadernowski, 1981). The process was performed using a hydrocyclone for oil cake and extraction meal. Hexane was used as a solvent. Meal containing 45% protein and 5–8% of crude cellulose was obtained with a 66% yield. The flour fraction can be used for enriching the meal in protein for swine and poultry feeding.

Meal from solvent extraction of the 00 Start rapeseed variety has been used instead of soyabean meal in mash for broiler feeding. The results obtained indicate that the presence of 17–24% of rapeseed meal does not reduce the slaughter value (Konarkowski and Konieczna, 1983).

2.1.2 Lecithin

2.1.2.1 Composition and properties

Degumming yields post-hydration gums which can be used as wastes or by-products. In the former case they are added to meal, while in the latter they serve for the production of lecithin. In the case of soyabean oil the processing of gums into lecithin is fully justified; however, in the case of rapeseed it can be inadvisable due to its lower price as compared with soyabean lecithin, which is due to poorer sensory properties. The choice of method of utilization of the gums depends therefore on the relation of the increased price of meal enriched with gums to the price of lecithin reduced by its production costs. The product obtained by drying gums, sometimes bleached, has a commercial name of lecithin. However, it is not pure lecithin, i.e. phosphatidyl choline, but it contains also ca. 14% phosphatidyl ethanolamine, 10% phosphatidyl inositol, as well as phytoglycolipids, carbohydrates, oil and water (Brekke, 1980). Chemical composition of the three basic constituents of commercial lecithin is presented in the figure:

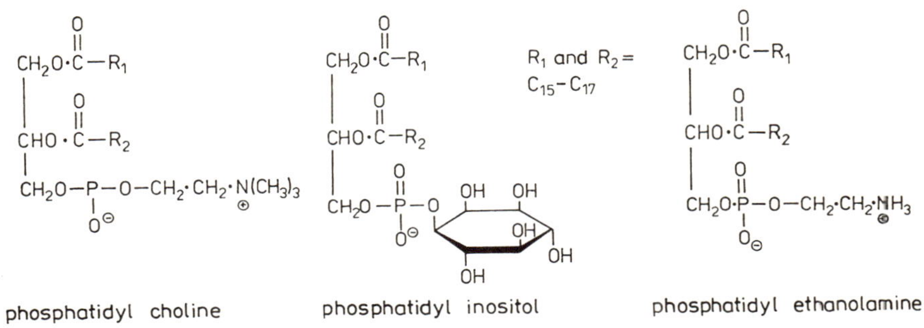

R_1 and R_2 = C_{15}–C_{17}

phosphatidyl choline phosphatidyl inositol phosphatidyl ethanolamine

Phosphatidyl choline and phosphatidyl ethanolamine have both cathionic and anionic character, hence they form zwitter ions, while phosphatidyl inositol is quite a strong acid. Lecithin containing all these three compounds has wetting and emulsifying properties. It forms a monomolecular layer at the oil/water interface, the fatty acid chains stretching into the oil phase, while the phosphoric acid or amine group orientate to the water phase. Such an orientation reduces the surface tension, enabling wetting, reducing the viscosity and stabilizing emulsion or disperse systems. Due to these, lecithin finds a widespread use as a wetting and emulsifying agent.

Soyabean lecithin is soluble in aliphatic and aromatic hydrocarbons, partly soluble in ethyl alcohol, and almost insoluble in acetone and water. Acetone dissolves in lecithin, the absorption continuing as long as the amount of acetone is insufficient for the precipitation of phospholipids. Lecithin is soluble in fatty acids, but almost insoluble in cold oils and animal fats. When heated it melts with them or forms disperse systems. After adding water to oil, lecithin undergoes hydration, forming a thick, yellow emulsion, which can be diluted with water. It is odourless and has a mild taste when properly produced. Hydration causes precipitation of not only phospholipids, but also other compounds and metals, which are subsequently contained in lecithin. The following compounds belong to this group: FFA, free and esterified oligosaccharides, glycolipids, sterols and their esters, tocopherols and carotenoids, Fe and Cu. Concentration of the metals in lecithin is 20–100 times higher than in the oil from which the lecithin has been obtained (Pardun, 1982).

The colour of unbleached soyabean lecithin changes from light yellow to brown, depending on the process parameters. Dry lecithin has a high viscosity, which first increases with an increase in the water content, and

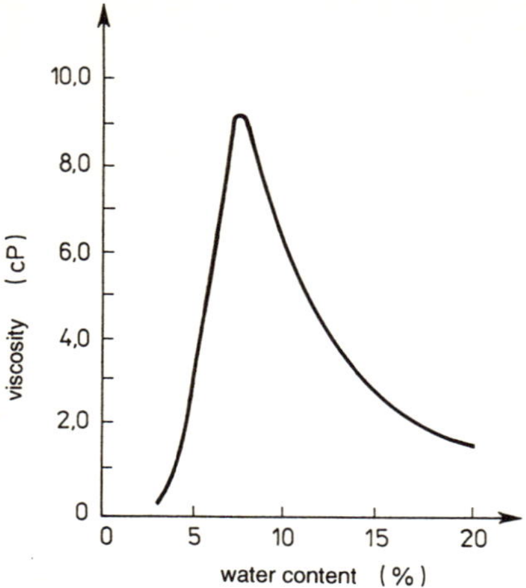

Fig. 2.1 Viscosity of lecithin vs. water content (Brekke, 1980).

then decreases, according to Fig. 2.1. The consistence of dry, plastic soyabean lecithin can be changed into fluid by an addition of 2–5% of soyabean oil fatty acids or their methyl esters. In such a form lecithin is more convenient to handle and dissolves easier in various solvents. Quality standards for soyabean lecithin are partly listed in Table 2.5; they comprise 6 grades. Lecithin is divided into plastic and liquid, each of these groups comprising unbleached, bleached and double bleached grades. Bleaching is usually performed with hydrogen or benzoyl peroxide, introduced either during hydration, or to a special apparatus (bleaching kettle). Refined lecithin is unstable, due to which it is mixed 1:1 with refined oil or sometimes with cacao butter. It is used for cosmetic purposes.

Apart from these, the following kinds of lecithin are also commercially available: mixed natural, deoiled, fractionated deoiled and chemically modified. Deoiled lecithin is obtained by extraction with acetone, which removes oil and fatty acids down to a 3–4% residue. It is next fractionated with alcohol, yielding an alcohol-soluble fraction — rich in phosphatidyl

Table 2.5 Quality standards for commercial soyabean lecithin in the USA (Brekke, 1980)

Components, properties	Content (%) in lecithin	
	fluid	plastic
Acetone insolubles	> 62	> 65
Water	< 1	< 1
Benzene insolubles	< 0.3	< 0.3
AV	< 32	< 30
Colour according to Gardner	4–10*	4–10*
Viscosity at 25°C (P)	< 150	—
Penetration (mm)	—	< 22

*Depending on the bleaching degree.

choline, and an alcohol-insoluble one — rich in phosphatidyl inositol. The characteristics of these fractions are given in Table 2.6. Lecithin is modified by hydrogenation, hydroxylation, acetylation, sulphonation and halogenation. Enzymatic hydrolysis yields a product of an increased emulsifying and dispersing power in aqueous systems. Hydroxylated lecithin is one of the more important products, since in contrast to natural lecithin it is readily dispersed in water and reveals greater emulsifying power towards fats.

Table 2.6 Characteristics of the particular fractions of refined commercial lecithin (Brekke, 1980)

Fraction components Emulsion type	Content (%) in fraction		
	deoiled lecithin	alcohol soluble fraction	alcohol insoluble fraction
Phosphatidyl choline	29	60	4
Cephalin	29	30	29
Inositol, other phospholipids and glycolipids	32	2	55
Soyabean oil	3	4	4
Others	—	4	8
Type of the emulsion formed	o/w or w/o	o/w	w/o

Rapeseed lecithin is considered inferior to soyabean lecithin, due to which it is usually not used as an emulsifier in edible fat technology. The main reason for its unpleasant smell and taste is autoxidation, taking place during separation and processing. The intensity of this reaction depends on the temperature of the processes the seed is subject to before the solvent extraction, and next the temperature of miscella distillation, hydration and drying of gums. Decomposition of the unstable phospholipid-protein complexes takes place during these processes, both directly and indirectly — by acceleration of autoxidation of fatty acids contained in phospholipid particles or oil accompanying them. The presence of water in the system also adversely affects the quality of lecithin (Niewiadomski *et al.*, 1972). The proper parameters of extraction and processing to eliminate these adverse effects have been proposed (Niewiadomski, 1970). The effect of drying the seed and of the solvent extraction (Sawicki *et al.*, 1975), as well as of drying the gums (Bratkowska and Niewiadomski, 1976), has been studied in detail.

Since the main fractions of commercial lecithin, i.e. cephalin and lecithin, stabilize emulsions of the opposite types, investigations have been performed on the possibility of obtaining emulsifiers for the oil/water systems (Sawicki and Niewiadomski, 1972). The characteristics of the commercial lecithin depends to some extent on the erucic acid content of rapeseed (Table 2.7).

Table 2.7 Characteristics of commercial lecithin obtained from rapeseed oil (Pardun, 1982)

Components, properties	Content (%) in lecithin from	
	zero erucic acid rapeseed	high erucic acid rapeseed
Water	0.18	0.39 ± 0.3
Acetone insolubles	71.5	71.1 ± 4.4
Phosphorus	2.25	2.23 ± 0.1
Lecithin	15.0	14 ± 0.3
Cephalin	7.4	7.6 ± 1
Lysolecithin	0.9	2.0 ± 0.4
Lecithin : cephalin ratio	2.0	1.9 ± 0.2
Colour in iodine scale	74	73 ± 24
AV	30	30 ± 1.6
IV	82.5	85 ± 3.3

2.1.2.2 *Utilization of lecithin*

Lecithin is utilized in the food industry, in pharmacy and in many other industries. In the food industry it is used as an emulsifier, wetting agent, and antioxidant. World annual demand for lecithin is estimated at 100,000 t; 95% of soyabean lecithin is used as an emulsifier in the food and fodder industries.

Lecithin not only plays the role of a raw material for various industries, but is also a synergist in many reactions. The proportion of lecithin used is of 0.1–0.3%. This is quite little compared to other components, but in numerous cases it is advantageous due to the fact that its smell cannot be detected. The taste in such amounts is also hardly perceptible, and for particular purposes specially refined lecithin or its fraction can be used. Lecithin is used in confectionery, e.g. for doughnuts, and in baking — to improve the absorption of water. Lecithin additionally increases the volume, prolongs the freshness and improves the homogeneity of bakings. In some products of the baking industry lecithin allows reducing the amount of fat without deteriorating the quality.

Unbleached liquid lecithin is most often used. It is dark brown, and therefore cannot be utilized in greater amounts for products of very light colours. In some countries the use of bleached lecithin is forbidden due to the presence of the residues of the bleaching agents.

Apart from the above applications, lecithin is used for the following purposes (Scocca, 1976):

— Some fodders, like milk substitutes for feeding calves, or other products of this kind, requring large amounts of fat. Lecithin facilitates the digestion of fat, stabilizes the product and acts as antioxidant.
— Besides the application in the production of chocolate, lecithin is used in the production of chewing gums, ice creams, pastes and other confectionery products.

—In margarine lecithin is used as an anti-sputtering and emulsifying agent. It yields a similar effect in the manufacture of shortenings and cooking fats. A particular role is played by lecithin in the recipes of dietetic and low-salt margarines, as well as mayonnaise. Lecithin reveals the best anti-sputtering properties when the ratio phosphatidyl choline : cephalin is 4.2:1.0. The emulsifying properties of lecithin depend on the manner of degumming. They are the best in the case of hydration with water, poorer in the case of application of citric acid, as well as acetic and maleic anhydrides, and the worst when phosphoric and oxalic acids are used (Smiles *et al.*, 1989).
—Lecithin secures solubility and dispersion of instant food products.
—As a flour additive lecithin improves kneading of dough and increases water absorption, also in the production of noodles.

Apart from the above applications lecithin is used in various pharmaceutical preparations and foodstuffs. Special varieties are used in the cosmetics industry for the production of derivatives by sulphonation, hydrogenation, etc., used as wetting and emulsifying agents. Lecithin finds also a widespread use as a stabilizing and dispersing agent in the production of gas oils, lubricants, as well as in rubber, soap and paint and varnish industries.

The use of soyabean lecithin is more and more widespread. At the same time the range of its varieties also extends. New methods yielding a product of the following properties have been developed (Bystram, 1980):

viscosity at 25°C	< 150 P — solid state
colour in the iodine scale	< 25–45
IV	< 26–35
PV	< 10–50
acetone insolubles	> 60–64

Fractionated lecithin of various phospholipid content is available.

Similarly to soyabean lecithin, sunflower lecithin can also be utilized in mash for laying hens and broilers. The results obtained indicate that a 2% lecithin addition increases the laying hens mass by 7.6% (Hollo *et al.*, 1985).

Cotton seed of the gossypol-free variety is the source of an oil, whose phospholipids can be utilized in a form of lecithin. This improves the profitability of the production, reduces the costs of wastes removal and decreases the possibility of emulsification during neutralization. These phospholipids contain only 0.23% of linolenic acid, hence there is no possibility of rapid oxidation and rancidity, typical for soyabean lecithin. Cotton seed lecithin contains also more lysolecithins, which improves its value as an emulsifying agent for foodstuffs (Cherry, 1983).

Rapeseed lecithin can play a similar role to soyabean lecithin provided that a proper technology is observed. Complaints concerning the quality of rapeseed lecithin initiated extensive investigations on its composition. They comprised not only lecithin, but also phospholipid gums from which it is derived (Niewiadomski and Stołyhwo, 1966; Zając and Niewiadomski, 1975).

In particular the composition of the fatty acids of the oil and of its phospholipids have been examined (Drozdowski and Niewiadomski, 1970). Another direction of the research focussed on the general composition of the phospholipids (Stołyhwo and Niewiadomski, 1974). Since NPL play an important role in refining of the oil, some of the investigations were aimed at their separation from HPL (Bratkowska and Niewiadomski, 1977) and at the determination of the composition of the hydrating fraction (Sawicki *et al.*, 1974). The results of the investigations on the effect of the particular unit operations on the quality of commercially available lecithin allowed the determination of the conditions necessary for the improvement of its properties (Niewiadomski, 1974). Gums from hydration of rapeseed oil mixed with meal reduce its dusting. Addition of 2% of gums to rapeseed meal does not affect the increase in mass of chicken and turkey, and probably larger additions can be used (Slinger, 1977). It has also been established that gums are not responsible for the hydrolysis of glucosinolates and are not goitrogenic. Therefore it can be generally stated that gums added to poultry rations in economically justified amounts induce neither beneficial, nor detrimental effects.

Other Canadian investigations (March, 1977) on the usefulness of lecithin obtained from 00 rapeseed revealed that an addition of 4% of gums to poultry rations yielded the same mass increase as in the case of addition of the same amount of soyabean oil. It has been demonstrated in another experiment that introduction of 5% of gums from both high- and low-erucic rapeseed together with 5% of maize oil to the diet yielded similar mass increase as an addition of 10% of maize or rapeseed oil.

Saponification of post-hydration gums with NaOH yields soaps, which are extracted with acetone after drying. Sterols are produced from this extract by crystallization. The yield is equal to 80–90% of the sterol content in this raw material. This method is not economical due to the price of lecithin (Sawicki, 1968).

2.1.3 Soapstock and post-refining acids

2.1.3.1 Composition and properties

The composition of soapstock depends on the type of neutralization equipment, kind, excess and concentration of the neutralizing agent, as well as the quality and characteristics of crude and neutralized oil.

Soapstock from batch neutralization contains usually up to 40%, and sometimes 50% of fatty acids (in total). On the other hand, soapstock from continuous neutralization contains 35–40% of the total fatty acids. The ratio of neutral fat to fatty acids in the form of soaps is equal in this case to 2:5.

A smaller soap content indicates errors during neutralization or the ocurrence of emulsification. A number of methods have been proposed to reduce the losses on alkali neutralization, viz. Zenith, Clayton, Halvopon, miscella neutralization, etc.

The composition of acidified soapstock depends on the ratio of the neutral oil to free alkali in the crude state. Its amount is usually sufficient to saponify nearly all the neutral fat in the soapstock. Saponified and acidified soapstock, containing 95% of fatty matter, usually contains 80%, and sometimes even 85–90% of free fatty acids.

Fatty matter of acidified soyabean oil soapstock has the IV of 125–135 and contains 75–85% FFA and 1–2% oxyacids. According to international trade practice, soapstock is sold on the basis of 50% fatty matter content. Soapstock always contains neutral fat in emulsified form. Sterols and tocopherols are present in soapstock, but their amount is lower in the deodorizer scum. To get the best price it is desirable that soapstock contains little pigments, neutral fat and other impurities (Balazs, 1987).

2.1.3.2 Utilization of soapstock

Basically crude soapstock is used for the production of soaps, and in acidified form — as fatty acids — is distilled. After distillation it is also used to a large extent for soap production. In cases when the fatty acids from this source contain large amounts of neutral fat before distillation, they are subject to hydrolysis. Distillation removes the characteristic smell of the acids better than saponification in a soap boiler. Bleaching is also improved.

Palm oil soapstock can be utilized directly for the production of low-quality soaps, particulary when the soap works are so close that soapstock can be pumped through a pipe-line. The soapstock can be also split yielding palm fatty acids. However, the process should be carried out directly after separation, since (especially in a warm climate) soapstock undergoes a rapid fermentation and emulsification. Soapstock splitting is carried out at 110–130°C, using a diluted mineral acid of pH 2.0–3.5. The water content in the fatty acids obtained in this way is 1–2%. According to trade practice, these acids should contain 95% fatty matter. When this content falls below 85%, the product can be rejected.

The soapstock from refining of lauric oil is usually sufficiently light and pure to be used directly for the production of certain kinds of soaps. Purification of soapstock consists in boiling with a small excess of lye in order to complete the saponification and subsequent graining out in the form of neat soap. This neat soap, containing ca. 60% fatty acids, is mixed with a small amount of water and is used in further processes for the production of soap. Except for lauric oils, soapstocks from other oils are not used for the production of soaps. This is due to a high degree of unsaturation, low congeal point and dark colour. However, such soapstocks can be used for this purpose after hydrogenation or can be used for feeding purposes and for the production of oleochemicals.

Soapstock in various forms can be used for animal feeding. When it is mixed with meal, its pH should be 5–7. It can be transformed into calcium salts and mixed in a powdered form with fodders.

Soapstock from 00 Tower rapeseed variety was added in an amount of 3.6–9% to a typical diet based on maize or soya in order to establish what would be the results compared to a similar diet containing the same amount of tallow (Summers, 1980). A diet containing an addition of a 1:1 mixture of soapstock and tallow was also examined. No differences in weight gain have been established between these three variants. Soapstock from the 00 variety can therefore be used for chicken feeding. It has been established in another investigation that a 1:1 mixture of these two fats increases the metabolized energy by 5.6% (Mutzar *et al.*, 1980).

Wastes formed during refining of fats contain sterols among others. The possibilities of their isolation and utilization have been examined on an example of rapeseed oil. Soapstock and oiled bleaching earth are particularly rich sources of sterols (Niewiadomski and Kłopotek, 1960). However, the condition of neutralization and bleaching cause transformations of sterols occurring in natural oils — formation of oxidation products and other derivatives in particular (Niewiadomski and Sawicki, 1958, 1964).

Dried soapstock contains ca. 4% water and is liquefied on heating to 45°C; even at room temperature it has a waxy consistence (Beal *et al.*, 1972). After hydrolysis of glycerides this product can be utilized as a raw material for the production of fatty acids, yet it finds the main use as an additive for poultry fodders, rich — in the case of soyabean soapstock — in xanthophyll. Beal *et al.* (1972) give the following composition of soapstock dried in this way:

	%		μg/g
total fatty acids	71.0–64.0	carotenes	350–575
FFA	34.0–28.3	xanthophyll	202–213
water	4.6–4.4		
UM	2.0		
phosphorus	0–1.4		
sodium	1.8–1.6		

It follows from the above results that soapstock processed in such a way maintains its natural carotene and xanthophyll content. A comparison of weight gains of broilers fed with a typical diet containing 4% of fat and 4% of neutralized, dried soapstock, did not reveal any differences. At the same time liquid wastes from the refinery did not contain the acidic, harmful component, viz. water separated after acidification of soapstock. A method has been patented in the USA for obtaining post-refining acids at a concentration exceeding 90%, reducing the viscosity of soapstock and thorough separation of the fatty acids from water (Red and Ilagen, 1978).

Soyabean oil soapstock can be utilized for the production of adducts of fatty acids with ethylene and butane. These compounds are used in cases when the fatty acids and their derivatives are not optimal for certain industrial applications. For example, saturated and unsaturated amines of C_{20} adduct have beeen applied as additives for polyethylene coatings.

2.1.4 Fats from adsorbents used for bleaching

2.1.4.1 Composition and properties

The fat content in the residue after boiling the spent bleaching earth in an autoclave is usually 6–8%. An improved method allows reducing this content to 2–3% (Swern, 1964). Oil recovered from bleaching earth is dark and contaminated as a result of intense oxidation induced by blowing the cakes in filter presses. Its quality can therefore be improved by reducing the blowing. In continuous bleaching it is recommended to recover the oil by solvent extraction. Processing of bleaching earth should be carried out as soon as possible to avoid storage, since oil contained in the earth rapidly decays, the AV rapidly increases, IV and yield decrease, while the polymer content increases.

2.1.4.2 Utilization of spent bleaching earth

Utilization of spent earth in chicken feed gave good results. Assuming that the methabolized energy of 1 kg of vegetable oil is equal to 37.7 MJ, it seems interesting to obtain 34.3 or 31.2 MJ from 1 kg of soyabean oil or palm oil contained in spent earth. However, the polymer content, equal for soyabean oil to 17.3%, raises some objections. Oiled earth can be used as a fodder additive provided that threshold value of mineral substances in feeding mixes is not exceeded.

Post-refining wastes, soapstock and bleaching earth from soyabean oil refining can contain such amounts of pesticides that their use for feeding purposes becomes impossible.

Bleaching earth from decolorization of sunflower oil, containing 30% fat, has an energy value of 12.5 MJ/kg. Experiments performed on broilers proved that 4% of the earth can be added to fodders, the fat contained in the earth being utilized in the same way as pure oil (Hollo *et al.*, 1985).

Granulation of sunflower meal accompanied by an introduction of oiled bleaching earth enriches the meal in the fat, sterols, tocopherols and microelements absorbed during bleaching. The addition (of 40–45% fat content) should be such that the meal contained 2.5–3.5% of fat at 9–11% humidity. The earth plays an additional role of a plasticizer during granulation (Šamraj *et al.*, 1980).

Bleaching earth oiled with high-erucic rapesead oil was used in an amount of 10 and 20% as an additive for broiler feed during 40 days. The weight gains were higher than in the control group. However, differences in the composition of the fat have been established. A 20% addition of bleaching earth caused an increase in the monounsaturated acid content at the cost of saturated and polyunsaturated acids (Łysakowski *et al.*, 1983).

2.1.5 Deodorizer scum

Distillate formed during deodorization is sometimes transferred to a skimming tank, from which the upper layer passes to sewer grease together with autoclave oil. This method eliminates the possibility of utilization of a few valuable components of crude oil, removed during deodorization.

2.1.5.1 Composition and properties

The chemical composition of the deodorizer scum depends on a number of parameters. First of all it depends on the type of the deodorized oil. For instance, palm oils yield condensates rich in FFA compared to rapeseed oil, as a result of the volatility of acids usually contained in the glycerides. The UM content also differs considerably, which is illustrated in Table 2.8. Age and quality (state) of the oil also influence the composition of the scum. Rancid oils are richer in oxidation products, which partly pass to the condensate. Deodorization parameters, particularly temperature, pressure and heating time, also play an important role (Niewiadomski and Drozdowski, 1962). The manner of condensation and the characteristics of water supplying the condenser also exert an influence.

The composition of deodorizer scum is so far not well known. Soyabean and cotton-seed oil scums basically have IV equal to 40–50, 33–45% UM, 14–28% sterols and their esters, as well as 9–15% tocopherols (Sawicki, 1968).

Investigations on the composition of wastes from rapeseed oil deodorization in a semi-continuous industrial apparatus revealed large differences in the composition of fatty acids occurring in the different fractions of the scum. For instance, the erucic acid content in the neutral oil was 46.7%, in FFA — 18.8%, in non-polar esters 23.2%, while in polar esters it was 51.2%. The presence of 60.5% of sterols, 11% of tocopherols and 27% of

Table 2.8 Composition of the fatty phase of deodorizer scum (Jakubowski and Przybyłowicz, 1963)

Components	Content (%) in condensate from deodorization of	
	coconut oil	rapeseed oil
Neutral fat	56–67	87–93
FFA	5.3–10.5	2.4–6.3
UM	28.0–34.0	2.0–7.2

hydrocarbons has been established in the unsaponifiable matter (Sawicki and Niewiadomski, 1970).

The wastes contain calcium and magnesium salts, which are formed as a result of contacting the fatty acids with water. Moreover, depending on the technological processes and the degree of fat decay, hydrocarbons, ketones and aldehydes can also occur. The distillate contains flavour compounds as well. In palm oil they are mainly ketones of high molecular weight, occurring in crude oil in an amount smaller than 0.1%.

Physical refining results in a decomposition of carotenoids, carbohydrates, protein compounds, peroxides and aldehydes. Only a small amount of sterols pass to the scum. The scum is a rich source of tocopherols and tocotrienols. Their content is equal to 3–8%.

Palm oil distillate is formed mainly during physical neutralization and — to a lesser degree — during traditional deodorization. Recently the production of palm oil fatty acids from alkali neutralization is decreasing, while the supply of the distillate from physical refining is increasing.

Table 2.9 Composition and chemical properties of distillate from physical refining of palm oil (Wong, 1983)

Components	Content (%)	FA composition (by GLC method)	
		acid	% w/w
FFA	85–90	$C_{12:0}$	0.3
SM	95–98	$C_{14:0}$	1.4
Water	0.1–0.5	$C_{16:0}$	47.4
UM	0.5–3.0	$C_{18:0}$	4.4
		$C_{18:1}$	36.9
Properties		$C_{18:2}$	8.4
		$C_{18:3}$	0.4
AV	51–55	$C_{20:0}$	0.4
c.p. (titer) (°C)	37–44	$C_{20:1}$	0.2

The characteristics of this distillate is listed in Table 2.9.

2.1.5.2 Utilization of deodorizer scum

In the sixties and the beginning of the seventies the installations for deodorizer scum recovery served not only for the protection of the enviroment, but were also a source of raw materials for the production of tocopherols and sterols. Tocopherols were used for the production of vitamin E, while sterols were used for pharmaceutical preparations. However, the interest in this source of tocopherols decreased in the years 1974/75, when synthetic vitamin E appeared.

The scum of high volatile UM content can be steam distilled and the fatty fraction can be utilized in the soap industry (Jakubowski and Przybyłowicz, 1963). However, when the UM content is lower than 5%, the processing is unprofitable, and it is utilized for feeding purposes. Sterols can be used for the production of steroid hormones, as oil/water emulsion stabilizers, etc. As a concentrate of vitamin E, tocopherols are particularly used for animal feeding. Ketones present in some condensates are used in the perfume industry. Distillate from physical refining can be used for the production of cheap soaps. Since it contains little water, it does not undergo fermentation or emulsification. However, as it is corrosive, it requires corrosion resistant equipment.

2.2 NON-EDIBLE FAT PROCESSING INDUSTRY

The non-edible fat processing industry, dealing with large amounts of fatty by-products from other industries, at the same time produces its own by-products. The most significant for the economy is glycerine — a by-product of all the processes of conversion of fats into fatty acids or their derivatives. Other by-products, like distillation residue or forerunnings, are less significant, mainly due to their smaller amount. However, sometimes very valuable, specific products, e.g. sterols, can be obtained from these raw materials also.

2.2.1 Glycerine

Aqueous solutions of glycerine, utilized for the production of pure glycerine or its concentrated solutions (70–98%), are formed as a by-product of all the processes of conversion of fats into fatty acid derivatives.

2.2.1.1 Sweet waters

Sweet waters obtained from non-catalytic splitting of fats contain 8–20% of glycerine on average, depending on the type of raw material being split. Splitting of post-refining acids, which can contain less than 50% neutral fat, yields sweet waters of a concentration even lower than 5%. Utilization of these waters depends on economical aspects. Sweet waters contain large amounts of impurities — water soluble organic substances, like amino acids and proteins from the split raw material, water soluble fatty acids, fat oxidation products, fatty acid salts and inorganic salts. The characteristics

Table 2.10 Average composition and reaction of sweet waters from non-catalytic splitting of animal fats and rapeseed oil (Mazgajska and Smenda, 1978)

Components, properties	Content (%) in product from	
	animal fats	rapeseed oil
Glycerine	13–18	15–20
Organics non-volatile at 160°C	0.05–0.5	0.05–0.6
Ash	0.05–0.2	0.05–0.2
pH	5.4–6.0	5.4–6.0

of sweet waters from non-catalytic pressure splitting of animal and vegetable fats is given in Table 2.10.

The quality of sweet waters, the manner of extracting pure glycerine from them and the yield of the process, depend primarily on the purity of the split raw material and the splitting degree. Refining should be carried out in the case of processing waste fats in order to remove protein and phospholipid impurities, as well as organic and inorganic salts of Fe, Na, Mg, Ca and Cr prior to splitting. Some of these compounds and incompletely split glycerides, i.e. mono- and diglycerides, possess emulsifying properties. In such cases a part of sweet waters remains in SFA, while a part of SFA contaminates sweet waters. According to Irodov (1976), an average water content of SFA should not exceed 0.3–0.5%. Under such circumstances the losses of glycerine in SFA do not exceed 1% of its initial content in fat.

The fat splitting degree should not be lower than 96–97%. Irodov (1976) recommends in the case of batch splitting washing the acids after the second splitting using water under high pressure. This allows increasing the splitting degree by 1–1.5%, since it prevents renewed esterification of FA with glycerine after reducing the pressure. Losses resulting from this esterification are equal to 0.7–0.8% when the splitting degree is 94–95%, and less than 0.5% when it is 96–97%. Sweet waters are a good medium for microorganisms, due to which they readily undergo decomposition during storage. Glycerine is obtained from sweet waters by removing organic and inorganic impurities and subsequent evaporation of water. Depending on the technology used, further losses of glycerine can occur during purification and removal of water as a result of either thermal polymerization, or steam distillation. When glycerine is obtained from sweet waters by chemical purification, evaporation of water and final refining by distillation, a waste is formed, viz. the distillation residue. Chemical purification of water allows removing most of the organic impurities, but it increases the ash content (Bełdowicz *et al.*, 1979). A high ash content of the crude glycerine, i.e. concentrated sweet waters directed to distillation, promotes polymerization during the distillation and increases the amount of the distillation residue.

Foaming is often observed during distillation of crude glycerine. It is caused by the decomposition products of compounds containing amine groups. They are mainly amino alcohols formed as a result of total or partial hydrolysis of phosphatides under the conditions of non-catalytic fat splitting. Water and high temperature also cause hydrolysis of proteins yielding amino acids. Darkening of sweet waters during storage is caused by reactions of glycerine oxidation products — aldehydes and ketones — with organic (mainly amine) compounds. Methods of chemical purification do not remove these compounds from sweet waters (Table 2.11).

Purification in an electric field is more efficient. According to Kamyšan *et al.* (1971), the electric field causes breaking of the emulsion and separation of the fraction initially dispersed in the water. This results mainly in a decrease in the lipid impurities content — 9 to 50 times.

Table 2.11 Amine nitrogen content expressed as mg alanine per dm^3 of sweet waters (Kamyšan and Janova, 1970)

Kind of sample	Amine nitrogen in sample No.					
	1	2	3	4	5	6
Water after splitting	540	446	510	645	687	836
Water after adding aluminium sulphate	528	426	490	629	671	824
Water after adding aluminium sulphate and next lime milk	500	416	450	605	639	794
Water purified in electric field	217	249	251	309	349	401

Ion exchange is often used for the purification of sweet waters. In this method sweet waters are freed from organic impurities by ion exchange in columns with anion and cation exchangers. Removal of impurities is so efficient that distillation of glycerine becomes unnecessary — it is sufficient to evaporate water down to the required concentration (Broniarz *et al.*, 1979). Owing to the elimination of glycerine distillation at high temperatures no polyglycerines are formed, hence no distillation residue occurs in this process. However, large amounts of liquid wastes are formed during regeneration of the ion exchange resins.

2.2.1.2 Soap lyes

Soap lyes are formed as a by-product during processing of neutral fats into sodium or potassium soaps. The lye is an alkaline aqueous glycerine solution contaminated with large amounts of salts and soaps. The lyes additionally contain organic impurities from the fat used for saponification (proteins, phospholipids). Neutral fats are most often used for the production of toilet soaps — they must be therefore light in colour and very pure. In the case of lower quality fats, soaps are usually made by neutralization of fatty acids obtained from them. Lyes of high content of organic impurities and low glycerine content are obtained when soapstocks are processed into sodium salts by further saponification. The characteristics of such lyes can for example be the following (Adamjan and Prokopeva, 1976):

Components	Percentage (%)
glycerine	2.5
soaps	2.1
chlorides	2.1
ether solubles	0.3
Properties	
pH	3.3
colour	dark
smell	soapy

Utilization of these lyes for glycerine production is energy consuming, but they cannot be directly disposed of as wastes since this would create a hazard for the environment, due to much too high a concentration of glycerine for biological water treatment plants, the allowable value being 500 mg/l.

An average composition and reaction of soap lyes from the production of toilet soaps is the following:

Components/properties	Content (%)
glycerine	6.2–8.8
ash	9.1–16.4
organics non-volatile at 160°C	0.8–1.3
ether solubles	0.3
pH	7–10

Lyes are characterized by a high salt content and strongly alkaline reaction. To obtain good quality glycerine from soap lyes it is necessary to perform the purification process properly. Maintaining strictly defined pH values at the particular stages is crucial.

After decomposing the soaps contained in the lyes using mineral acids, the fatty acids are separated, the lyes are again alkalized to pH 8.0–8.3 and aluminium sulphate is added until pH 4.5 is reached. Aluminium soaps are separated by filtration and lye is again alkalized to pH 10.5–10.8 in order to bind the excess of aluminium sulphate into insoluble hydroxide separated by filtration. Too strong an alkalization causes partial dissolution of aluminium hydroxide in alkalis, thus increasing the ash content in lyes (Turanskiy *et al.*, 1971). A waste — aluminium soap deposit from filtration — is formed during purification of soap lyes. Crude, so-called lye glycerine can be obtained from soap lyes. The ash content of this glycerine is higher than in saponification glycerine obtained from sweet waters. Further processing into high quality glycerine is the same for both glycerines.

2.2.1.3 Other by-products containing glycerine

Transesterification of fats with low molecular weight alcohols yields apart from methyl or ethyl esters, also a by-product — waste containing glycerine of various concentration and various purity, depending on the transesterification technology applied.

Glycerine by-product obtained from transesterification of rapeseed oil with methanol has the following characteristics (Jaworski, 1978):

Components	Percentage in esterification product	
	batch method	continuous method
glycerine	51.2	30
ash	6.6	1
non-volatile organics	6.6	0.6

Properties		
alkalinity (% w/w Na_2O)	2.1	—
pH	—	7–8
n_D	1.4280	—
density at 20°C (kg/m^3)	1.141	—
viscosity at 20°C (cP)	48.6	—

A waste is obtained in the process of reduction of glycerides to fatty alcohols. It contains mainly glycerine decomposition products, i.e. a mixture of propylene glycols and propyl alcohol (Hinze, 1984), since the process is carried out at a high temperature and under high pressure.

Ammonolysis of triglycerides (Billenstein and Blaschke, 1984) yields aqueous glycerine solutions of relatively high purity as a by-product.

Glycerine produced from the by-products discussed constituted in 1982 74% of total world annual production (420,000 t in 1982). Only 26% of this valuable chemical was obtained by synthesis. It is anticipated that the production of glycerine from natural sources will further increase (Heidrich, 1984).

2.2.2 Distillation residue

Distillation residue is a by-product from the purification of fatty acids or their derivatives by vacuum distillation at an elevated temperature. The distillation residue contains mainly the less volatile or non-distilling components of the raw material, as well as oxidation and polymerization products, i.e. high molecular weight products formed under the influence of high temperature. Usually, independently of the raw material from which it originates, the distillation residue is dark and has therefore a limited applicability. The amount of distillation residue can be estimated at 5–10% of the mass of the distilled material on average. Such amount is therefore significant, due to which much attention is paid to methods for its rational utilization.

2.2.2.1 Composition and properties

Depending on the kind of the processed raw material the distillation residues differ significantly both with respect to colour and the possibilites of utilization.

The greatest amount of distillation residue is formed during FA distillation. The so-called first residue from animal SFA distillation (13–20% relative to the mass of the raw material) contains mainly unsplit glycerides, a part of FA and UM (Table 2.12). It can be split again and distilled, the process yielding the so-called second residue amounting to 5–10% of the mass of the initial SFA. The characteristics of the second residue differs largely from the first residue as a result of further thermal changes of the fatty matter. The acid number of the second residue should not exceed 50. Fatty acids obtained from distillation of the split first residue are usually of poorer quality than the distillate from SFA.

Table 2.12 Characteristics of the first residue from distillation of fatty acids obtained from animal fat splitting (Szczepańska *et al.*, 1971)

Components, properties	Residue after distillation of fatty acids from		
	tallow	utilization grease	bone fat
AV	19–112	47	100–118
SV	131–196	196	186–204
IV	56–66	58	49–55
Self-ignition temp. (°C)	370–383		
		%	
Fatty matter	98–98.2	—	—
UM	1.5–11.7	2.4	2.3–3.1
Ash	0.6–2.2	—	—
Oxyacids	—	1.5	—

The distillation residue obtained after a repeated splitting and distillation of tallow methyl esters can be enriched in unsaponifiable matter up to 10%. The UM contains 30–50% of sterols (Struve and Schuh, 1985).

The distillation residue from distillation of unsaturated acids, particularly those with conjugated bonds (e.g. dehydrated castor oil acids), contains large amounts of polymers. Under the influence of high temperature a part of the dienes polymerize, increasing the amount of the non-volatile by-product even up to 40%. The characteristics of such a distillation residue of dehydrated castor oil is the following (Leščenko and Levit, 1973):

Property	Value	FA composition	% w/w
AV	128.5	C_{16}	1.6
SV	203.7	C_{18}	0.8
IV	123.6	$C_{18:1}$	8.5
UM (%)	0.6	$C_{18:2}$ isolated	35.7
		cis-trans and *cis-cis* conjugated	39.6
		trans-trans conjugated	18.5

The residue from tall oil distillation, so-called tall oil pitch (TOP), contains large amounts of UM, vegetable sterols included. An example of the characteristics of TOP is given by Holmbom (1978):

Components	Content (%)
Total free acids	34.6–51.6
fatty	0.8–2.4
resin	3.3–12.5
other	25.7–42.1
Total esterified acids	23.3–37.8
fatty	8.2–15.2
resin	0.9–1.9
other	8.8–23.5
Total UM	25.3–30.1
low molecular	14.1–17.7
high molecular	9.6–15.5

The following acids have been found in the group of free acids: oleic, pimaric, isopimaric, palustric, dehydroabietic and abietic. In the group of esterified acids — palmitic, linoleic and *trans-trans* conjugated linoleic, as well as dehydroabietic. Diterpene alcohols, pimarol, docosanol, tetracosanol and β-sitosterols, campesterol, etc., have been found in the UM.

The residue from distillation of castor oil methyl esters pyrolyzed in order to obtain oenanthal is a thick, oil liquid of AV 12–28 and the viscosity of ca. 55 cP at room temperature; it contains (Walisiewicz-Niedbalska and Jaworski, 1977):

	%
Polymerized acids	up to 30%
Esters of acid	
ricinoleic	18–27
linoleic	6–36
oleic	14–27
palmitic	5–22
undecylenic	3–18

2.2.2.2 *Utilization of distillation residue*

Distillation residues containing large amounts of sterols (sitosterols from TOP, and cholesterol from tallow acids distillation residue) can be utilized as sources of these substances.

Struwe and Schuh (1985) recommend a repeated splitting and distillation of the residue until the sterol concentration reaches 10%. The remaining fatty acids are separated by distillation after methanolysis, the process being followed by saponification of UM. Cholesterol can be used as a raw material for the production of progesterone.

Sterols can be isolated from the TOP by a method similar to that for sulphate soaps, but the yield is lower. For this purpose the saponified residue is dissolved in ethyl alcohol, extracted and refined by crystallization, absorption and other methods. TOP can be also utilized for the production of thermoplastic glues. A reaction with isocyanates yields coating-forming polyurethanes (Surewicz and Surma-Ślusarska, 1983).

Asphalt emulsions were obtained from the second residue from animal fat splitting. About 30% of the residue, 30% of asphalt, machine oil and concentrated ammonia forming the emulsifier (ammonia soaps) were emulsified with water and sodium metasilicate. Stable oil/water emulsions serving as anti-corosive metal coatings were obtained (Szczepańska *et al.*, 1971).

FA distillation residue modified with ethanolamine can be used as an additive for asphalts, improving the adhesion of asphalt to aggregate (Pilichowski and Kajl, 1973).

Vegetable and animal FA distillation residue can be converted into sodium soaps; after mixing with naphthenic acid soaps a flotation agent for the enrichment of phosphorites is obtained. Application of 0.91 kg of this agent per 1 t of rock allows extraction of ca. 82% of P_2O_5 (Pokrowskij and Santalov, 1971).

2.2.3 Forerunnings

Forerunnings are fractions of the lowest boiling point collected during distillation refining of fatty acids or their derivatives. Apart from the main component, i.e. fatty acids or their derivatives, depending on the kind of the raw material, they contain also volatile UM and low-molecular weight FA. The composition of forerunnings from distillation of animal and vegetable FA in Lurgi equipment is given in Table 2.13.

Table 2.13 Characteristics of forerunnings from distillation of animal and vegetable fatty acids (Cegłowska *et al.*, 1972)

Components, properties	Forerunnings from distillation of		
	tallow FA	waste animal FA	vegetable FA
IV	34.4–36.2	37.1	31.6–37.2
PV	1.4–2.7	1.0	0.0–1.2
		%	
Fatty matter	97.5		97.9–98
UM	0.5–1.1	0.6	1.8–2.2
Oxyacids	0.5–1.0	0.9	0.2–0.8
Water	0.01–0.1	—	
		mg/kg	
Sulphur	6.2–12.1	9.8	13.7–22.2
Iron	0.9–4.8	12.7	2.8–41.7

Composition of FA			
Acid	% w/w		
Below C_8			0.4–0.9
C_8	0.6–2.0	0.7–0.9	5.1–10.6
C_{10}	1.4–2.6	1.2–1.3	4.5–5.1
C_{12}	3.5–5.0	1.8–2.0	13.5–25.0
C_{14}	9.5–12.4	8.2–9.6	6.8–7.6
$C_{14:1}$	1.7–2.9	1.8–2.0	0.6–0.9
$C_{14:2}$	1.0–1.4	0.8–1.0	0.3–0.4
C_{16}	32.7–38.9	36.2–37.6	16.9–28.8
$C_{16:1}$	5.7–6.5	5.4–6.0	2.3–4.0
$C_{16:2}$	0.9–1.2	1.2–1.5	0.5–0.6
C_{17}	0.8–1.0	0.8–1.1	0.3–0.6
C_{18}	6.5–8.1	7.6–9.4	4.3–7.6
$C_{18:1}$	21.2–25.9	25.8–26.9	14.6–20.3
$C_{18:2}$	3.1–4.1	3.7–4.3	3.7–6.3
$C_{18:3}$ + C_{20}	0.0–0.6	0.0–1.2	0.7–1.6
$C_{20:1}$	—	—	0.5–1.1

Forerunnings from distillation of tall oil, so-called tall light oil (TLO), are obtained with ca. 10% yield relative to the crude oil (Holbom, 1978). TLO contains ca. 32% volatile UM and only ca. 2% resin acids. The SV is ca. 140, while the AV — 137. Linoleic, oleic and palmitic acids are the main components of FA. Volatile UM consists mainly of hydrocarbons of molecular weight 256–258 and resin acid decarboxylation products.

Forerunnings are utilized for flotation of iron ores. Palmitic acid used for the production of various derivatives can be isolated from animal FA and TLO.

2.2.4 Monomers

Polymerization of fatty acids or their methyl esters yields — apart from the polymerization product, i.e. acid or ester dimers — distilled monomers as a by-product. Their composition is slightly different from that of the raw material used for the polymerization and also depends on the technology used. Monomers are removed from the main product by distillation. They are light in colour and can be used instead of olein for the production of many preparations.

The monomers — by-products of dimerization oleine catalyzed by clay — have the following characteristics (Rybczyńska *et al.*, 1973):

Congeal point (°C)	28–32
AV	166–168
SV	168–171
IV	25–44
UM (%)	5–10

The fatty acids comprise odd and even carbon number, saturated and unsaturated C_7—C_{20} FA.

The properties and the composition of monomers — by-products of dimerization of low-erucic rapeseed oil methyl esters, catalyzed by boron fluoride phenolic complex, are the following (Chmielarz *et al.*, 1984):

Properties		FA composition	% w/w
IV	73	C_{16}	9.8
AV	12	C_{18}	3.7
colour in iodine scale	1	$C_{18:1}$	66.7
viscosity at 20°C (cP)	10	$C_{18:2}$	8.2
		C_{20}	2.2
		$C_{20:1}$	6.0
		$C_{20:2}$	3.4

2.2.5 Lanolin soaps

Alkali neutralization of suint, carried out in order to obtain lanolin, yields a by-product — so-called lanolin soaps. These soaps contain ca. 50% water, fatty acid sodium soaps, waxes and free alcohols. The content of free alcohols in lanolin soaps can be 8–20% (Walisiewicz-Niedbalska *et al.*, 1985). Lanolin soaps contain also impurities from the crude suint.

Lanolin soaps do not find direct applications. However, post-refining lanolin acids obtained by acidification of lanolin soaps can be used e.g. as a valuable lubricant additive. These acids have AV of ca. 62, SV ca. 120, and contain up to 4% of water.

2.3 OTHER INDUSTRIES

Non-edible fats of great economic significance are formed as by-products during processing of many raw materials (Section 1.3). They constitute a large part of the fatty raw materials utilized for the production of washing agents and various fatty derivatives.

2.3.1 Slaughter and poultry fats

The characteristics of slaughter fats depend on a number of factors — first of all on the origin of the tissue used for rendering (Table 2.14), and on

Table 2.14 Melting point and iodine value of pork fat depending on its origin (Rutkowski, 1974)

Adipose tissue	m.p. (°C)	IV
Perinteric, large intestine	40–43	44–54
Perinteric, small intestine	41–44	40–50
Perigastric	40–42	44–45
Perirenal	39–41	48–54
Back fat	37–38	58–59
Peripancreatic	42–45	42–48

the manner and time of storage of the adipose tissue before rendering. For instance, pork tissue fat changes its properties during storage at 4°C in the following way (Rutkowski, 1974):

Tissue	AV	PV
fresh	0.4–1.1	0.1–0.3
after 2 days	0.2–2.1	0.2–0.4
after 4 days	0.7–3.6	0.4–0.6
after 6 days	1.8–8.0	0.5–0.7

Slaughter fats are by-products of a relatively high purity and with a low degree of triglycerides decomposition. Animal fats from meat processing industry are esters of saturated C_{16} and C_{18} acids, as well as unsaturated $C_{16:1}$ and $C_{18:1}$ acids. They are used mainly for the production of DFA utilized for the production of amines, sodium soaps, stearin and olein (particularly soft fat). They can also be directly saponified in order to obtain toilet soaps, domestic hard soaps or powders.

Table 2.15 Characteristics of poultry fat (Szczepańska *et al.*, 1979)

Components, properties	Content (%) in poultry fat from	
	solvent extraction	aqueous extraction
Ether solubles	92.6	92.0
Water and volatile substances	6.0	5.9
Ether insolubles	0.2	0.8
Protein	1.0	0.6
Bound glycerine	8.8	9.4
AV	21	16.4
SV	185	194
PV	1.0	27.0
IV	77.0	78.2
Colour in iodine scale	636	170
FA composition (by GLC method)		
Acid	% w/w	
C_{14}	0.8	0.8
C_{15}	0.1	0.2
C_{16}	21.8	21.3
C_{18}	5.8	4.6
C_{20}	1.8	1.4
$C_{14:1}$	0.4	0.4
$C_{16:1}$	9.5	11.5
$C_{18:1}$	40.1	41.3
$C_{20:1}$	0.4	0.1
$C_{18:2}$	16.5	16.8
$C_{18:3}$	1.2	0.8

Competitive utilization of these raw materials by the fodder industry is limited by a low content of linoleic acid and high AV. According to Terpstra *et al.* (1978) low FFA contents are insignificant with respect to the nutritive value; assimilation of fat starts to decrease only when it contains more than 50% FFA. During digestion the triglycerides are partly decomposed to monoglycerides, which are emulsifiers. At high FFA contents the amount of emulsifier formed is too low.

The characteristics of poultry fat are given in Table 2.15. Poultry fat is usually processed into DFA, and further — e.g. to technical olein, since it contains large amounts of oleic acid. Due to a dark colour and relatively high contamination with decomposition products it cannot be utilized directly for feeding purposes.

2.3.2 Fish oils

The composition of crude fish oils can vary significantly depending on the kind of the raw material processed, as well as the time and manner of storage. Fish oils of a quality unfit for consumption, and fish post-refining acids of the characteristics listed in Table 2.16 are used for technical

Table 2.16 Characteristics of fish oils and post-refining acids (Czapiga and Mazgajska, 1982)

Components, properties	Content (%) in	
	crude fish oil	fish post-refining acids
Ether solubles	97.3–98.6	89.9–97.6
UM	1.3–2.0	1.9–3.5
Protein	0.14–0.15	0.56
Water	0.8–1.0	1.1–6.7
AV	7.6–18.3	99.8–106.8
SV	191–193	192.6–193.5
IV	134–160	160.4–162.4
PV	18.9–19.7	1.7–1.9
Colour in iodine scale	100–170	1076
	FA composition (by GLC method)	
Acid	% w/w	
C_{14}	3.1–4.4	4.0–4.4
$C_{14:1}$	0.4–0.7	0.4–0.5
$C_{14:2}$	0.4–0.6	0.3–0.4
C_{16}	12.7–16.5	11.7–13.3
$C_{16:1}$	16.7–20.1	12.5–13.0
$C_{16:2}$	1.6–2.2	0.9–1.0
$C_{16:3}$	1.1	0.3
C_{18}	2.7–3.1	2.6
$C_{18:1}$	23.5–30.4	28.9–29.6
$C_{18:2}$	3.3–5.4	5.2–6.0
$C_{18:3}$	0.8–3.1	3.0–6.6
$C_{20:1}$	4.6–7.8	5.4–5.9

purposes. Owing to their high IV they can be used after refining for the production of protective coatings, lubricants and tannery oiling agents (Niewiadomski, 1984).

2.3.3 Utilization grease, degras, bone fat

Average characteristics of these raw materials are given in Table 2.17. A characteristic feature of all these products is a very diversified quality, both with respect to the degree of glyceride decomposition, and the oxidation degree and the content of impurities.

The production of utilization grease is largest. This grease contains the largest amount of fatty matter of a relatively small extent of decomposition, and therefore after a proper refining it can be used for feeding purposes. Depending on the technology of leather processing and the kind of the raw material, the degras can contain large amounts of oxidation products even after a direct rendering (Table 2.18). Due to this it can be used only for technical purposes after splitting and distillation.

Bone fat obtained from fresh bones can even be used for direct consumption, since it contains essential fatty acids (linoleic), has low melting

Table 2.17 Characteristics of utilization grease, bone fat and degras (Cygańska and Bełdowicz, 1971)

Components, properties	Utilization grease	Bone fat	Degras
		%	
Ether solubles	96.5–98.5	89.1–98.6	81.6–92.3
UM	0.8–1.3	0.3–0.5	0.4–2.6
Oxyacids	0.98–1.57	1.0–1.7	0.9–1.5
Water	0.1–0.4	0.6–1.1	0.2–1.2
Ash	0.01–0.17	0.01–1.08	0.21–0.84
Protein	0.14–1.69	0.09–0.34	0.34–1.31
		mg/kg	
Phosphorus	11.2–59.7	8.5–55.2	0.0–16.4
c.p. of acids (°C)	36.0–39.9	34.4–40.4	34.6–39.2
m.p. (°C)	34.4–37.8	33.6–44.3	36.2–37.7
SV	179–197	183–198	184–201
AV	11.2–41.3	33.5–104	44.5–153.5
IV	39.1–59.8	42.2–53.7	41.2–59.7
PV	2.4–18.0	9.5–19.2	9.3–15.1
Colour in iodine scale	47–1076	100–1076	100–1076
		FA composition (by GLC method)	
Acid		% w/w	
C_{12}—C_{14}	2.5–5.8	1.6–4.0	1.6–3.2
C_{16}	25.4–28.8	25.1–30.6	24.5–32.6
$C_{16:1,2,3}$	5.3–9.6	4.4–5.2	3.5–7.8
C_{18}	10.1–14.3	15.6–17.3	7.5–17.7
$C_{18:1}$	38.2–47.2	43.0–46.5	41.2–53.5
$C_{18:2}$	1.6–7.2	0.0–6.2	1.1–4.1

Table 2.18 Yield and characteristics of degras (Rutkowski, 1974)

Composition, properties	Kind of raw material			
	machine shavings		dry separated pork shavings	pork shavings after tanning and washing
	beef	pork		
		Content (%)		
Water	24–35	28–33	5.7–8.7	68–71
Fat	37–62	59–65	7–36.8	13–20.5
Protein	5–14	2.3–3.2	0.5–1	6–12
Ash	0.1–5	1–5	58–66	4–7.5
Yield of rendering (%)	70–85	62–94	61–75	44–66
		Properties of rendered fat		
m.p. (°C)	31–44.5	25–28	24–28	25–33
AV	2.8–6.7	1.5–1.9	1.3–7.5	2.6–11.7
PV	0.7–6.5	2–12.5	8.8–12.6	4.5–10.7
IV	42–54	61–62	59.5–62.7	52.6–59.4
SV	189–211	179–257	147–202	201–210
UM (%)	0.5–7.4	0.4–2.6	0.4–1.4	0.6–4.2

point and is readily emulsified. Liberman *et al.* (1974) propose to add it to cooking fats in an amount of 10–20% instead of tallow. Bone fat from stored bones can be used only for technical purposes. It is a good raw material for the production of olein and stearin.

2.3.4 Tall oil

The content of resin acids in tall oils of various origin can vary within a 20–60% range, the content of FA — 18–70%, and UM — 5–24%. A complete characteristic of tall oils produced throughout the world falls within the following limits (Kocór and Kroszczyński, 1978):

Properties		Components	Content (%)
density (kg/m^3)	0.95–1.025	ash	0.4–0.6
viscosity (cP)		water	0.4–7.0
at 18°C	$760–15\times 10^4$	petroleum ether	
at 100°C	150–1200	insolubles	0.1–8.5
AV	107–174		
IV	35–216		

The following average properties and composition of tall oil from pine wood processing (Surewicz and Surma-Ślusarska, 1983) can be taken as typical:

AV	166–170
IV	170–175

Component	Content (%)
resin acid	42–51
FA	41–51
UM	7

Composition of UM

behenyl and lignoceric alcohols
aliphatic and terpene hydrocarbons
phytosterols: β-sitosterol, campesterol, dihydrositosterol

Composition of FA	%
oleic and linoleic	80–90
linolenic, palmitic, stearic, lignoceric, behenic, lauric, myristic, arachidonic	10–20

Composition of resin acids	%
abietic, neoabietic, palustric, dextropimaric, isodextropimaric	50–75
abietic acid disproportionation products: dehydroabietic, dihydroabietic, tetrahydroabietic	25–50

Crude tall oil finds limited application. It is usually refined by distillation and fractionated in order to isolate resin acids used for the production of paper glues, rosin and fatty acids, e.g. for the preventive coating industry. Due to the presence (in an amount of 9–13%) of linoleic acid with conjugated bonds, and the lack of linolenic acid, tall oil fatty acids are particularly suited for the production of yellowing-resistant protective coatings.

2.3.5 Suints

Suint is a wax, i.e. a mixture of esters of aliphatic alcohols and sterols with fatty acids. Suint contains also free fatty acids, free alcohols and hydrocarbons. The alcoholic part constitues ca. 40–50% w/w, 20% of which are aliphatic alcohols of straight and branched iso- and ante-iso- chains, ca. 29% cholesterol, and the rest — triterpene alcohols and diols. The acid part contains (Szczepańska and Szelejewski, 1979):

	%
C_{10}—C_{26} n-acids	7
C_{10}—C_{28} iso-acids	22
C_{10}—C_{28} ante-iso-acids	30
Hydroxyacids	30

The remaining part of suint consists of both organic and inorganic substances, compounds of sulphur and certain metals included (Busujuščij *et al.*, 1970).

Owing to the presence of a great number of components of specific and seldom encountered composition, suint is a valuable source of pharmaceutical and cosmetic preparations. It is generally utilized in two ways:

1. Without refining — as a lubricant additive, hence for a typically technical purpose.
2. After refining — as lanolin, or after fractionation or chemical modification — as lanolin derivatives. In pharmacy and cosmetics it is a source of emulsifiers and improvers of specific and irreplaceable action.

2.3.6 Furred animal fats

Furred animal fats are relatively pure products. Rendered under preservative conditions they are light in colour, have a slight smell and the consistency of lard. After additional refining or fractionation into solid and liquid fractions (at room temperature) they can be used as components of the fatty phase in the production of cosmetic emulsions. Furred animals fats can be also used for synthesis of cosmetic emulsifiers (Kosko, 1983).

REFERENCES

Adamjan R.J. and Prokopeva M.F. (1976), *Masložir. Prom.*, **10**, 26.
Balazs I. (1987), *J. Am. Oil Chem. Soc.*, **64**, 1126.
Beal R. *et al.* (1972), *J. Am. Oil Chem. Soc.*, **49**, 447.
Bell J. and Jeffers H. (1976), *Can. J. Anim. Sci.*, **56**, 269.

Bełdowicz M. *et al.* (1979), *X Sesja Naukowa Kom. Technol. Żywn. PAN*, Kraków, abstracts, p. 255.
Billenstein S. and Blaschke G. (1984), *J. Am. Oil Chem. Soc.*, **61**, 353.
Bragg D. (1974), *Canola Council of Canada*, **35**, 106.
Bratkowska I. and Niewiadomski H. (1976), *Acta Aliment. Pol.*, **26**, 251
Bratkowska I. and Niewiadomski H. (1977), *Acta Aliment. Pol.*, **27**, 39.
Brekke O. (1980), In: *Handbook of Soy Oil Processing and Utilization*, Erickson D. *et al.* (Eds.) Am. Soyabean Ass., St. Louis and Am. Oil Chem. Soc., Champaign, p. 71.
Broniarz J. *et al.* (1979), *X Sesja Naukowa Kom. Technol. Żywn. PAN*, Kraków, abstacts, p. 458.
Busujuščij V. N. *et al.* (1970), *Masložir. Prom.*, **10**, 19.
Bystram K. (1980), *Biul. Inform. Inst. Przem. Mięsnego i Tłuszczowego*, **18** (3), 44.
Cegłowska K. *et al.* (1972), unpublished work.
Cherry J. (1983), *J. Am. Oil Chem. Soc.*, **60**, 360.
Chmielarz B. *et al.* (1984), unpublished work.
Clandinin D. *et al.* (1978), *Rapeseed Assoc. Canada Publ.*, **51**, 1981, Canola Council of Canada Publ., 59.
Crauer L. (1970), *J. Am. Oil Chem. Soc.*, **47**, 210A.
Cygańska J. and Bełdowicz M., *Pollena*, **15**, 20.
Czapiga M. and Mazgajska I. (1982), unpublished work.
Daun K. *et al.* (1982), "Oilseeds-processing", In: *Grains and Oilseeds*, Canadian International Grains Institute, Winnipeg.
Drozdowski B and Niewiadomski H. (1970), *Zesz. Probl. Post. Nauk Roln.*, **91**, 101.
Heidrich J. F. (1984), *J. Am. Oil Chem. Soc.*, **61**, 271.
Hinze A .G. (1984), *Fette, Seifen, Anstrichm.*, **86**, 520.
Hollo J. *et al.* (1985), *Proc. ISF Congress in New Dehli.*
Hussar A. (1977), *Rapeseed Assoc. of Canada*, **45**, 137.
Holmbom B. (1978), *J. Am. Oil Chem. Soc.*, **55**, 342.
Irodov M. W. (1976), *Masložir. Prom.*, **12**, 19.
Jakubowski A. and Przybyłowicz S. (1963), *Tłuszcze i Środki Piorące*, **6**, 128.
Kamyšan M. and Janova L.I. (1970), *Masložir. Prom.*, **9**, 17.
Kamyšan M. A. *et al.* (1971), *Masložir. Prom.*, **8**, 23
Kocór M. and Kroszczyński W. (1978), unpublished work.
Konarkowski A. and Konieczna L. (1983), *XIV Sesja Naukowa Kom. Technol. Żywn. PAN*, Poznań, abstracts, p.118.
Kosko I. (1983), *Zesz. Probl. Post. Nauk Rol.*, **32**, 147.
Leščenko P. S. and Levit M.S. (1973), *Masložir. Prom.*, **5**, 28.
Liberman L. G. *et al.* (1974), *Masložir. Prom.*, **8**, 14.
Łysakowski K. *et al.* (1983), *XIV Sesja Naukowa Kom. Technol. Żywn. PAN*, Poznań, abstracts, p 16.
March B. (1977), *Rapeseed Ass. of Canada*, **50**, 95.
Mazgajska I. and Smenda J. (1978), unpublished work.
Mazgajska I. *et al.* (1982), unpublished work.
Minkowski K. *et al.* (1978), *Rocz. Inst. Przem. Mięsnego i Tłuszczowego*, **15**, 243.
Mutzar A. *et al.* (1980), *Canola Council of Canada*, **57**, 82.
Niewiadomski H. (1970), *Proc. of Intern. Conf. on the Science, Technol. and Marketing of Rapeseed*, Ste Adele, Quebec, abstracts, p. 223.
Niewiadomski H. (1984), *Fatty Raw Materials* (in Polish), WNT, Warszawa.
Niewiadomski H. and Drozdowski B. (1962), *Roczn. Techn. Chem. Żywn.*, **9**, 47.
Niewiadomski H. and Kłopotek A. (1960), *Roczn. Techn. Chem. Żywn.*, **5**, 91.
Niewiadomski H. and Sawicki J. (1958), *Grasas y Aceites*, **9**, 306.
Niewiadomski H. and Sawicki J. (1964), *Fette, Seifen, Anstrichm.*, **66**, 930.
Niewiadomski H. and Stołyhwo A. (1966), *Roczn. Techn. Chem. Żywn.*, **12**, 189.
Niewiadomski H. *et al.* (1972), *World Congress of the ISF*, abstracts, p. 72.
Pardun H. (1982), *Fette, Seifen, Anstrichm.*, **84**, 1.
Pilichowski B. and Kajl M. (1973), unpublished work.
Porkowskij V. N. and Santalov B.A. (1971), *Masložir. Prom.*, **7**, 41.
Red J. and Ilagen J. (1978), USA patent No. 4,118,407.
Roberts R. (1976), *J. Am. Oil Chem. Soc.*, **53**, 302.
Rutkowski A. (1974), unpublished work.
Rybczyńska B. *et al.* (1973), unpublished work.
Šamraj G.I. *et al.* (1980), *Masložir. Prom.*, **1**, 11.

Sawicki J. (1968), *Utilization of By-products from Vegetable Oil Refining* (in Polish), Wyd. Przem. Lekkiego i Spożywczego, Warszawa.
Sawicki J. and Niewiadomski H. (1970), *Zesz. Probl. Post. Nauk Roln.*, **91**, 273.
Sawicki J. and Niewiadomski H. (1972), *Roczn. Techn. Chem. Żywn.*, **22**, 39.
Sawicki J. *et al.* (1974), *V Sesja Naukowa Kom. Technol. Żywn. PAN*, Gdańsk, abstracts, p. 34.
Sawicki J. *et al.* (1975), *Acta Alimen. Pol.*, **25**, 323.
Scocca P. (1976), *J. Am. Oil Chem. Soc.*, **53**, 428.
Slinger S. (1977), *Canola Council of Canada*, **45**, 93.
Smiles A., Kakuda Y. and McDonald B. (1989), *J. Am. Oil Chem. Soc.*, **66**, 348.
Sosulski F. and Zadernowski R. (1981), *J. Am. Oil Chem. Soc.*, **58**, 96.
Stołyhwo A. and Niewiadomski H. (1974), *V Sesja Naukowa Kom. Technol. Żywn. PAN*, Gdańsk, abstracts, p. 28.
Struve A. and Schuh R. (1985), *Fette, Seifen, Anstrichm.*, **87**, 103.
Summers J. (1980), *Canola Council of Canada*, **57**, 80.
Surewicz W. and Surma-Ślusarska B. (1983), unpublished work.
Swern D. (1964), *Bailey's Industrial Oil and Fat Products*, Interscience Publ., New York.
Szczepańska H. and Szelejewski W. (1979), unpublished work.
Szczepańska H. *et al.* (1979), *Pollena*, **23**, 123.
Szczepańska H. *et al.* (1971), unpublished work.
Terpstra K. *et al.* (1978), unpublished work.
Turanskij G. Z. *et al.* (1971), *Masložir. Prom.*, **11**, 20.
Walisiewicz-Niedbalska W. and Jaworski A. (1977), unpublished work.
Walisiewicz-Niedbalska W. and Jaworski A. (1978), unpublished work.
Walisiewicz-Niedbalska W. *et al.* (1985), unpublished work.
Wong M. (1983), *J. Am. Oil Chem. Soc.*, **60**, 316.
Zając M. and Niewiadomski H. (1975), *Acta Aliment. Pol.*, **25**, 63.

CHAPTER 3

Composition, Properties and Utilization of Wastes

The wastes which can be utilized will be mainly discussed in this chapter. Therefore it does not comprise fumes, as well as liquid and solid wastes which pollute the environment and have little value as raw materials. Wastes are formed both in the edible oil industry and in the non-edible fat industry.

3.1 SEED IMPURITIES

The fraction of useful rapeseed impurities separated on a cleaner contains on average 95% of these impurities and 5% of whole seeds (Minkowski and Katzer, 1981). The impurities contain 7.7% water and 25.4% fat. Directly after isolation the oil from these impurities has the AV of 58.8; after 1 month it is 72.3, while after 2 months — 78.8.

The impurities, containing a larger amount of hydrophilic substances (proteins and sugars) compared to seeds, have a moisture content higher by an average of 1% during storage.

The increase in the AV depends on the damage of the rapeseed structure. The more it is damaged, the higher the hydrolysis degree. In seed halves it is lower. Solvent extraction of the impurities is faster than that of undamaged seed. Should the impurities be utilized as an oily raw material, the fraction should be processed as soon as possible after the extraction due to its instability. Impurities of this kind can be added to silaged maize green forage, since this causes a reduction of the amount of glucosinolate derivatives. On the other hand, the AV and PV increase (Kozłowski *et al.*, 1981).

The content of proteins, ash, cellulose, solubles and nitrogen-free substances, as well as fat in impurities from sunflower seed cleaning has been determined (Garbuzova, 1979). It has been established that 95% of samples from production contained from 10.12 to 38.32% fat. Impurities of high fat content can be utilized for the production of technical oil. Due to a high protein content these impurities can be used as fodder additives.

3.2 HULLS

The hulls play an important role in sunflower seed processing, since dehulling results in an increase in the protein content of meal to > 42% and reduces the cellulose content below 12%, which widens the possibilities of utilization of this meal as a fodder additive (Hollo *et al.*, 1985).

Dehulling reduces the volume of the pressed and solvent extracted mass, thus increasing the capacity. Assuming that sunflower seed contains 22–28% of hull and that 8% of it is left after dehulling, 1 t of seed yields more than 150 kg of hull.

Recent results indicate that the hull can be utilized in the folowing way (Bender, 1980):

- —for the production of phenolic glues used in plywood production,
- —as a component of moulding compositions during production of bakelite resins.

Moreover, rapeseed hulls can be utilized for some feeding purposes. When they are totally fat free, sheep digest 40–49% of them, while when they contain oil residues they are 56–61% digested (Michalet-Doreau *et al.*, 1983). They can be added in an amount of 25% to feed mixes for cattle, and 10% for rabbits (Baudet and Evrard, 1978).

Experiments on young fattened animals proved that an 18% addition of sunflower hull resulted in weight gains higher by 4.1–6.1% compared to the control diet (Hollo *et al.*, 1985).

Although rapeseed dehulling increases the nutritive value of meal, it simultaneously reduces its mass. Most probably the higher price does not compensate for the loss of mass. Nevertheless, research on dehulling is continued, since it is possible that rapeseed will become a raw material for the production of protein preparations for human nutrition.

Depending on the fat content, the sunflower hull has an energy value of 16,700–17,000 kJ/kg, which enables 5 t of steam to be obtained from 1 t of hulls. This amount covers the steam demand of an oil mill with a surplus. The ash formed in an amount of ca. 3% contains valuable components, like 26–28% K_2O or 7–7.5% P_2O_5, and therefore can be utilized as a fertilizer for various plants. Burning of sunflower seed hulls can be basically carried out in boilers of any design (Leibovitz and Ruckenstein, 1981). For instance, an oil mill processing daily 400 t sunflower seed yields 3.2–3.6 t hulls per hour, which corresponds to steam production of 10–12 t/h. Since only 6.5 t is necessary in the oil mill, the rest can be utilized in other parts of the plant.

Burning of sunflower seed hulls creates certain difficulties and an environmental problem (Müller, 1983). This is due to a rapid formation of ash layers on the boiler pipes, which hinders the heat transfer. To avoid this, it is necessary to modify the design of the water-tube boiler.

Hulls can be also processed into furfural. It is formed by hydrolysis of pentoses, xyloses and arabinoses, which are the components of various agricultural wastes, like straw, hull, etc. Sunflower seed hull contains 24–26% of pentoses, hence it yields 14.9–17.1% of furfural. The wastes from this production can be utilized as a fuel.

3.3 PALM OIL MILL WASTES

The use of dried mill effluent as a supplement for animal feeds has been investigated (Hutagalung, 1987). Satisfactory results were obtained using 15% in broiler rations, 10% for layers and pigs and 20–40% for cattle. The effluent contains 13% ether extract, 6.1% digestible protein and a gross metabolizable energy content of 7.7 MJ/kg for poultry, 9.2 for pigs and 9.8 for cattle.

In extensive trials with goats, good live weight gains were achieved when a mixed herbage ration of Napier grass and Leucaena was supplemented by up to 30 g dried effluent per kilogram live weight (Vadiveloo, 1988).

3.4 SPENT HYDROGENATION CATALYST

Nickel catalyst that lost its activity during hydrogenation can be used again after regeneration. The fat content of the spent catalyst is usually 40–50%, while the nickel content — 8–12%.

The regeneration is carried out by incineration and subsequent dissolution in nitric acid. Nickel oxide or hydroxide can be obtained from the nitrate solution after its purification. The activity of the regenerated catalyst depends among others on the reduction temperature. It has been established that the highest activity is achieved when the temperature is equal to 480°C in case of obtaining oxide and 290°C when obtaining hydroxide. Compared to a standard catalyst, whose activity is assumed to be 100, the catalyst regenerated from oxide reveals 94% activity determined by the decrease of the refractive index of a standard oil at 40°C, while that obtained from nickel hydroxide has 96% activity. No differences have been established in the duration of the period of activity of the regenerated catalyst compared to a standard catalyst (Massoumi and Kajbaf, 1979). Spent nickel catalyst is a waste that should not be disposed of in landfill (Hastert, 1989). It may be sold for recycling into nickel-containing manufactures. For this purpose it is important to use the minimum amount of filter aid when filtering out the nickel. Excess filter aid increases oil losses and transport costs and the consequent reduction in nickel concentration reduces the price obtained.

3.5 SPENT BLEACHING EARTH

Regeneration of bleaching earth has not been carried out until now, since the earth is inexpensive and most often there are no problems with disposing of it at waste dumps. Moreover, attempts to regenerate the spent bleaching earth revealed that it recovers less than 50% of the activity of fresh earth (Ong and Sinkeldam, 1983). However, oiled bleaching earth is used for certain purposes. Gočitašvili *et al.* (1977) used it as a component of cattle and poultry diet. The following compounds are adsorbed on bleaching earth during refining of fats and oils: vitamins, sterols, FFA, phospholipids, chlorophyll pigments and carotenoids. All of them are valuable fodder components. Spent bleaching earth has the consistency of a paste, which makes its introduction to fodders difficult; due to this it has been mixed with medium ground maize at a 1:2 ratio at a temperature of 60–70°C.

Spent bleaching earth, containing 25–40% fatty matter, can be processed into an emulsifier by an addition of monoethanolamine (Ibrachimov and Isajev, 1982). The emulsifier has the consistency of a thick paste of 55–60°C m.p., is brown and solidifies at 40–45°C. It contains 40–60% fatty acids monoethanolamides, 35–55% dry matter (bleaching earth) and 5% water at the most. It is poorly water soluble, but dissolves readily in petrol, toluene and petroleum derivatives.

Spent bleaching earth is used together with other oily wastes for the production of lubricants used for drawing of metals. According to লožešnik (1970) such a lubricant contains soap wastes from the refining of edible hydrogenated sunflower oil (8–10%), gums, fat from bleaching earth, and spent catalyst (10–15%), free alkalis (0.3–0.5%) and water. Among the numerous possibilities of utilization of spent bleaching earth, Fahn (1984) mentions its utilization in the building material industry.

3.6 FRYING OILS

Waste frying oil from cook shops has been added in an amount of 20% to mineral oil used in diesel engines. The engine revealed reduced fuel consumption and smoking compared to other additives and regular diesel fuel. The majority of vegetable oils have excellent ignitability under the operating conditions of diesel engines and can be utilized as improvers (Pryde, 1980).

3.7 SEDIMENTS AND LIQUID WASTES

Sediments accumulating in crude oil tanks are important oil industry wastes. Usually they are periodically removed and added to autoclave oil or to meal. The sediments from vegetable oil tanks are utilized for animal feeding mixes. Fats obtained during purification of liquid wastes can be utilized for feeding purposes provided that their residence time in sewage systems was short and their decomposition degree is not excessively high (Sawicki, 1983). Addition of fatty components to feed mixes plays an important role — 1 g of fat yields 37 J, while that of carbohydrates and protein — ca. 17 J. Moreover, fats are carriers of oil soluble vitamins and anti-oxidants. However, such utilization of fats is limited by their price. It should be taken into account that fats are 95% digested, while maize is 85% digested. The composition and properties of fats used for feeding are the following (Wong, 1983):

FFA	ca. 15%	c.p. of acids	44°C
water	0.5–2.0%	PV	15–20
impurities	traces	AOM stability test	ca. 20 h
UM	< 3.0%	colour	light

Materials used for oiling of meal should contain more than 82% fatty matter and less than 15% HCl-insoluble ash.

Wastes from settling tanks of fat processing plants have been successfully used as a component of mycelium substrate in the production of mould cheese (Sołtysek, 1983), for baker's yeast cultures (Kwaśnik, 1983) and for the production of citric acid (Elimer, 1983).

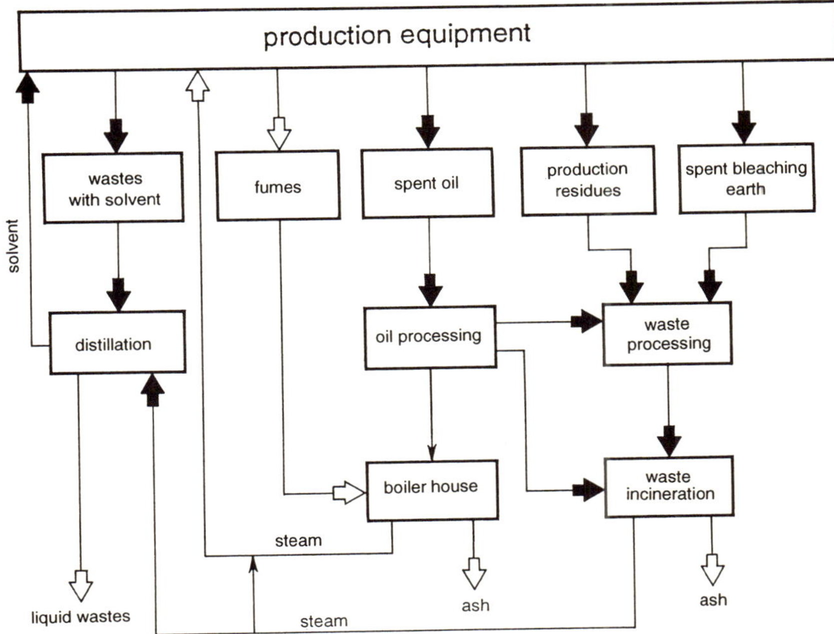

Fig. 3.1 Oleochemical plant waste management (Dieckelmann *et al.*, 1983).

3.8 WASTES FROM OLEOCHEMICALS PRODUCTION

Production of oleochemicals yields wastes in the form of fumes, solvent-containing liquid wastes, spent oils, production residues, as well as bleaching earth cakes and active carbon (Fig. 3.1).

Noxious wastes are formed during production of fatty acids. They can be removed by burning in a boiler. Filter presses, stirrers, skimming tanks, separators and tanks are connected to the ventilation system through drip tubes and coolers. Solid wastes and filter press cakes are processed into a pumpable material and fed to a special burner, where they are incinerated (Dieckelmann *et al.*, 1983). Exhaust gases are utilized in steam generators. Small amounts of ash and possibly silicates have to be removed.

Liquid and melted fatty wastes are heated, which enables the separation of water and sludge. The fatty substances obtained serve for maintaining the flame in the burner. They are supplied to the burner through a flow-through container with continuous stirring. An outline of a system of waste incineration is illustrated in Fig. 3.2. Fatty wastes are atomized in the burner by means of superheated steam and a rotating nozzle. The fumes serve as an oxygen source during burning. Steam for production purposes is obtained in a heat exchanger. The incinerated wastes consist mainly of fatty acids, fatty alcohols and esters of various chain lengths. Sludges from

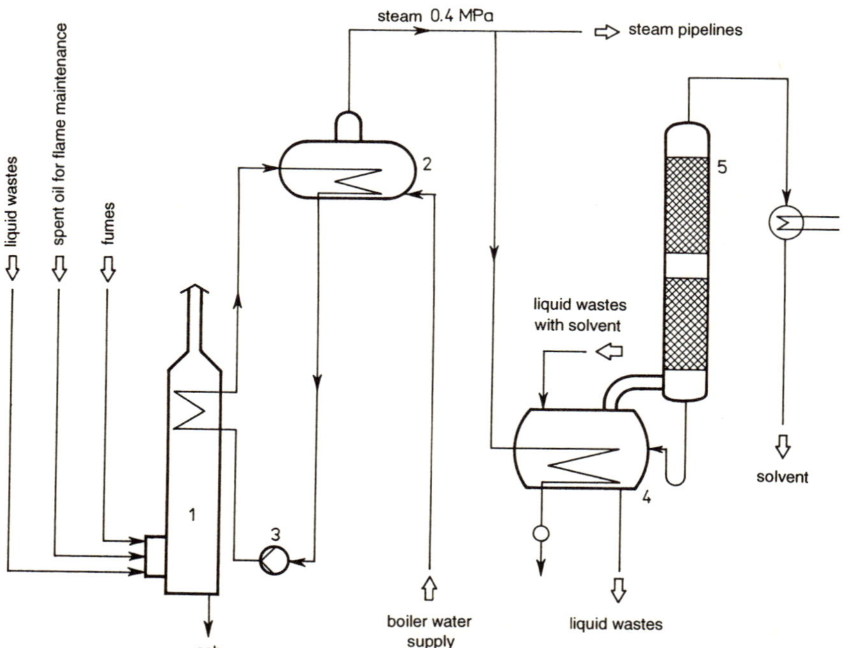

Fig. 3.2 Equipment for incineration of wastes in oleochemical plants (Dieckelmann *et al.*, 1983): 1—furnace, 2—steam generator, 3—heating oil circulation, 4—heating of wastes before distillation, 5—distiller for solvent-containing water.

presses have a paste consistency and contain 40–60% esters. Liquid wastes are kept at 60–80°C to avoid their solidification. In order to properly operate the system it is necessary that the waste are composed in such a way that they are pumpable and that their energy value does not drop below 3500 heat units.

3.9 WASTES FROM FISHING INDUSTRY

Fat-containing wastes are formed in all fish processing plants. They accumulate as a surface scum in skimming tanks in the form of a stable, dense emulsion containing 30–55% water, 40–60% fat, protein substances and various impurities soluble in both phases (Wocial and Usydus, 1984), as well as parts of fish and residues of packages. Depending on time and storage conditions, the fat contained in these scums undergoes rapid hydrolysis, oxidation and other reactions decreasing its usefulness. The FFA content increases by 50–90%. The scums are periodically disposed of by waste removal cars. The amount of these wastes is hard to estimate, even approximately; in big fish processing plants it is ca. 4 t per month. In cases when the residence time of the fats in the sewage system is shortened to 1–2 days, the wastes can be utilized for the production of gear oils and rolling greases.

Poorer kinds can be mixed with fuel oil. Separation of the fatty matter is performed by decantation or in a two-stage process comprising degumming and phase separation. In both cases it is necessary to heat the wastes to 90°C.

The fish processing industry, comprising the production of canned food, pickles, fillets, as well as salted, smoked and gutted fish and other products, produces wastes which sometimes contain large amounts of fat, for instance (Usydus and Suchy, 1979):

Liquid wastes	mg/dm^3
production of canned food	120 ± 286
production of pickles	1399 ± 2376
production of liver oil	25926 ± 16920
production of fish meal	12256 ± 30349

Fish oil sediments usually contain 40–60% of oil. They are processed into fatty acids.

3.10 OTHER WASTES

3.10.1 Glycerine distillation residue

Glycerine distillation residue contains mainly polyglycerines (20–50%), glycerine (up to 40%), fatty acids and nitrogen compounds. The chemical composition of the residue depends both on the technology of glycerine production and on the quality of the raw material. The residue, containing ca. 40% glycerine and other compounds with hydroxyl groups, yields after esterification with e.g. wastes formed during the production of sebacic acid, in the presence of sulphuric acid (5% w/w) as a catalyst at a temperature of 130–140°C, a product of the AV equal to 65–75, being a good grease for rolling of pipes. The presence of long-chain polyglycerines, well covering the metal surface and increasing the viscosity of the grease, means that the grease is better than that obtained e.g. by thermal concentration and saponification of wastes formed during the production of sebacic acid (Breskina *et al.*, 1978).

Another method of utilization of glycerine distillation residue consists in dilution with hot water (70–80°C) to a density of 1.25–1.32 and addition in an amount of 0.2–0.6% to casting cores, where it serves as a binding agent. Due to this the consumption of other binding agents is reduced to 1–1.5% (Panenceva *et al.*, 1976).

3.10.2 Wastes formed during the production of sebacic acid from castor oil

Wastes formed during the production of sebacic acid from castor oil contain (Tarasenko *et al.*, 1974; Leščenko *et al.*, 1984):

Acid	%
oleic	25–50
stearic	5–10
linoleic	15–20
dihydroxystearic	5–10

triglycerides and FA polymerization products	%
FFA	80
bound FA	ca. 10
UM	ca. 5
substances insoluble	
in acetone	25
in petroleum ether	ca. 20

This waste can be used for the production of cooling lubricants used for cold and hot forming of metals.

3.10.3 Wastes from meat processing plants amd mass nutrition utilities

Wastes from the production of lard have the BOD of 180 and COD of 220 (Howe, 1977).

Crude sewer grease from a skimming tank has the consistency of a thick sludge and is an emulsion of fat and a water-protein phase. Its quality cannot be standardized, sine its composition is very diversified. On the basis of 28 samples Rutkowski (1974) gives the following composition of sewer greases from meat processing plants and mass nutrition utilities:

	%
water	1–89.5
fat	5.8–98
proteins	0.2–6.0
total solid impurities	0.3–9.8
in this amount: organic	0.1–8.9
ash	0.05–3.1

Sewer greases are partly rendered in fatty acid production plants.

3.10.4 Miscellaneous

Lower qualities of tallow and greases, as well as palm-oil post-refining acids and condensate from physical refining, can be used for feeding purposes. The possible presence of sulphuric acid or the products of its reaction with other components decrease the palatability for the majority of animals. Similarly, the presence of pesticides renders utilization in such a way impossible.

Fatty acid distillation residue contains large amounts of UM and polymers; after physical refining—also UM, but small amount of polymers. The standard for waste oils for poultry feeding (Röttgermann, 1983) recommends m.p. below 30°C and the following composition:

	%
FFA	< 70
water and petroleum ether insoluble impurities	< 1.5
UM	< 6
linoleic acid	> 20
linolenic acid	< 10
erucic acid	< 2.5
polymers	< 5
monomers, unoxidized fatty acids	> 85

REFERENCES

Baudet J. J. and Evrard J. (1978), *Proc. 5th Inter. Rapeseed Conf.*, Malmöe, Vol. 2.

Bender F. (1980), *Research ond Canola seed, oil meal and meal fractions*, **57**, 257, Canola Council of Canada, Ottawa.

Breskina A. J. *et al.* (1978), *Masložir. Prom.*, **4**, 43.

Dieckelmann G. *et al.* (1983), *Fette, Seifen, Anstrichm.*, **85**, 559.

Elimer E. (1983), *XIV Sesja Naukowa Komitetu Techn. i Chem. Żywn. PAN*, Poznań, abstracts, p. 116.

Fahn R. (1984), *Fette, Seifen, Anstrichm.*, **86**, 505.

Garbuzova I. (1979), *Masložir. Prom.*, **9**, 21.

Gočitašvili K. R. *et al.* (1977), *Masložir. Prom.*, **7**, 41.

Hastert R. C. (1989), *J. Am. Oil Chem. Soc.*, **66**, 174–176.

Hollo J. *et al.* (1977), *Proc. ISF Congress in New Delhi.*
Howe R. (1977), *Proc. Meat Industry Research Conf.*, Moscow, abstracts, p. 97.
Hutagalung R. I. (1987), *Critical Reports on Applied Chemistry*, **15**, 85. Ed. F. D. Gunstone, Society of Chemical Industry, London.
Ibrachimov A. S. and Isajev Ch.J. (1982), *Maslozir. Prom.*, **11**, 40.
Kozłowski M. *et al.* (1981), *Zesz. Nauk. ART Olsztyn, Zootechnika.*
Kwaśnik I. (1983), *XIV Sesja Naukowa Komitetu Techn. i Chem. Żywn. PAN*, Poznań, abstracts, p. 106.
Leibowitz Z. and Ruckenstein C. (1981), *Fette, Seifen, Anstrichm.*, **83**, 534.
Leščenko Z. J. *et al.* (1984), *Maslozir. Prom.*, **11**, 32.
Massoumi A. and Kajbaf M. (1979), *J. Am. Oil Chem. Soc.*, **56**, 565.
Michalet-Doreau B. *et al.* (1983), *Abstracts 6 Intern. Repeseed Conf.*, Paris.
Mińkowski K. and Katzer A. (1981), *Rocz. Inst. Przem. Mięsnego i Tłuszczowego*, **17/18**, 123.
Muller S. (1983), *J. Am. Oil Chem. Soc.*, **60**, 393.
Ong J. and Sinkeldam E. (1983), *Fette, Seifen, Anstrichm.*, **85**, 304.
Panenceva N.N. *et al.* (1976), *Maslozir. Prom.*, **4**, 36.
Pryde E. (1980), in: *Handbook of Soy Oil Processing and Utilization*, Erickson D. *et al.*, Am. Soybean Ass., St. Louis and Am. Oil Chem. Soc., Champaign, p. 459.
Röttgermann P. (1983), *Fette, Seifen, Anstrichm.*, **85**, 190.
Rutkowski A. (1974), unpublished work.
Sawicki J. (1983), *XIV Sesja Naukowa Komitetu Techn. i Chem. Żywn. PAN*, Poznań, abstracts, p. 30.
Sołtysek K. (1983), *XIV Sesja Naukowa Komitetu Techn. i Chem. Żywn. PAN*, Poznań, abstracts, p. 70.
Tarasenko R. J. *et al.* (1974), *Maslozir. Prom.*, **7**, 33.
Usydus Z. and Suchy Z. (1979), unpublished work.
Vadiveloo J. (1988), *Proceedings of International Oil Palm Conference*, July 1987, p. 417, Palm Oil Research Institute of Malaysia, Kuala Lumpur.
Wocial M. and Usydus Z. (1984), unpublished work.
Wong M. (1983), *J. Am. Oil Chem. Soc.*, **60**, 316.

CHAPTER 4

Protection of the Natural Environment

Increasing pollution of the natural environment, particularly of water and the atmosphere, provokes a growing reaction of society in defence of human, animal and plant health and lives. The action has two directions: design and application of equipment to reduce the hazard, and utilization of repressive measures towards those responsible for the pollution. For example, in the seventies the USA assigned 2.2% of its national income for this purpose, and Germany 1.8% (Choffel, 1976). Protection of the environment against the detrimental effect of fats occuring as main or accompanying raw materials, as well as against other agents from the fat industry, comprises the problems of fumes, liquid wastes and solid wastes.

4.1 FUMES

Fumes from the fat industry contain dust, solvent vapours, fatty acids, aldehydes, ketones, other chemicals, and sometimes the odours of decaying raw materials, main and by-products, and all the improperly processed wastes. The fumes originate mainly from production halls and the equipment. Typical examples of air pollution comprise (Becker, 1972):

1. In the air surrounding the plant:
   —dust,
   —noxious odours,
   —results of failures, like fires, cracks, leakages of tanks, etc.
2. In the air from production halls and the equipment:
   —dust,
   —noxious fumes,
   —solvents,
   —chemicals used.
3. Air from venting the tanks and the stores.
4. Air from cooling towers:
   —chemicals,
   —oils,
   —fatty acids.

4.1.1 Dust

Dust in the air of production halls or stores creates a great explosion hazard. Bigger dust particles settle on the ground surrounding the plant or adjacent ground. During rain they are washed to the sewage system, which can increase the BOD.

Extraction of fats involves the use of large amounts of air, necessary for purification and drying of the seed before pressing and solvent extraction and for desolventizing the meal. In some cases air is necessary for dehulling. Particularly large amounts of air are required for drying the seed. When the raw material is not well purified, the stream of air removes bigger particles, for example hulls, during drying. This air is most often purified by means of self-cleaning sieve filters. Fine dust can be removed by means of cyclones and fabric filters. However, the air leaving the desolventizer is hot and damp, therefore it is impossible to pass it through such a filter. Application of a cyclone is sufficient due to the fact that humidity facilitates settling of the dust in the cyclone (Goodrich, 1980).

The following factors are sufficient to create a dust explosion hazard: the presence of dispersed dust in air in a concentration within its explosion limits, a sufficient amount of oxygen in the air and an ignition source. Lack of any of these three factors excludes the possibility of an explosion. Dustiness of the internal air during filling and emptying of silos cannot be avoided. On the other hand, thorough cleaning can eliminate this hazard outside the chamber.

The second factor is always present and cannot be avoided, unless an inert gas, e.g. CO_2 or nitrogen, is supplied to the bottom of the silo in such an amount that the concentration of oxygen drops below 8%. The third factor comprises sparking due to mechanical shocks, welding, electrostatic discharges, heating resulting from friction, self-ignition and failures of electric installations. Electrostatic discharges can be eliminated by connecting all the conducting elements and grounding with a conductor of resistance lower than $10^6\,\Omega$. Self-ignition can occur in cases of prolonged storage, hence it can be counteracted by controlling the temperature and periodically transferring the contents of the chamber (Zockoll, 1982).

The dust content of hot, comminuted soyabean is ca. 2%. About 1% originates from the first grinding and only this amount is transferred during dehulling. The second part comes from the second grinding. Compared to 5% of dust formed during cold comminution this means a significant reduction. The hot comminution system enables additionally a 50% reduction of the emission of air compared to the traditional system. The amount of wastes is therefore much smaller.

Dust explosion hazard occurs in the case of copra pressing when dust concentration exceeds 30 g/m^3 air (Zockoll, 1979).

4.1.2 Fumes from solvent extraction

Fumes from extraction of oil seeds originate from the processed raw material and from the solvent used. Fumes from rapeseed extraction can be an example of the former. The specific odours are not neutralized during solvent recovery and are released to the atmosphere. It has been established that they consist mainly of hydrogen sulphide and acetaldehyde. The problem is encountered during processing of both high erucic and low erucic rapeseed. In the former case the concentration of H_2S is 9.5 g/m^3, while that of acetaldehyde is 8.2 g/m^3; in the latter case the figures are 6.1 g/m^3 and 6.7 g/m^3, respectively. Hydrogen sulphide originates from the decomposition of cysteine and sinigrine, and acetaldehyde is formed during heating of the seed to 90°C. The two compounds are not only noxious, but also harmful. Hydrogen sulphide causes corrosion, while acetaldehyde is poisonous (Lindh and Dahlen, 1989).

Due to absorption in an absorption bulb filled with mineral oil, the solvent vapour content in air is usually so small that no other devices are necessary. The same concerns desolventizing of meal and oil.

Fumes from the desolventizer and final meal cooler can contain lots of dust, fibres, etc., particularly when the raw material is readily pulverizable. Various scrubbers, cyclones, etc., are used to stop the dust, which prevents fouling of condensers and other parts of the equipment for solvent recovery (Swern, 1964).

Air is supplied to the bottom of the chamber and passed through meal stored in the silo. In this way solvent residues are removed. On its way to the top of the meal column the air can even become saturated with solvent vapours before it leaves the chamber. Meal containing so much solvent releases the solvent to the surrounding atmosphere. The lower explosion limit of petrol in air is ca. 2.3%. The range of concentrations of mixtures exploding in a 19 mm diameter tube comprises 2.5–8% (Koziorowski, 1980). The explosion limits of a hexane-air mixture are 1.2–7.4% v/v (42–265 g/m^3). A hexane-air mixture cannot explode when the hexane partial pressure drops below 11 Pa. This takes place at temperatures lower than –28°C. The solvent content of meal is proportional to its fat content. Table 4.1 presents the dependence of the content of petrol in the gaseous phase contacting the stored meal on fat content of this meal for various temperatures. The gaseous phase content corresponds to the lower explosion limit of the mixture. It follows from this Table that on heating the meal containing 2.2% of oil from 20 to 40°C, the hexane content corresponding to the lower explosion limit of the gaseous mixture decreases from 0.05 to 0.03%. In order to maintain a safety margin the meal stored at 20°C should contain in the gaseous phase not more than 75% of the concentration of the solvent corresponding to the lower explosion limit. Data given in Table 4.1 can change from year to year, depending on the climate of the cultivation place. It has been established for instance that rapeseed collected in 1980 had different properties with respect to solvent desorption and

Table 4.1 Content of petrol in meal resulting in exceeding the explosion limit after equilibration (Radant, 1982)

Meal	Content in meal (%)				
	fat	hexane at temperature of			
		20°C	25°C	30°C	40°C
Soyabean, sunflower	1.0	0.03	0.028	0.027	0.025
	1.6	0.036	0.034	0.032	0.028
Coconut, palm kernel	2.0	0.042	0.037	0.035	0.03
	3.3	0.056	0.051	0.047	0.042
	4.7	0.07	0.068	0.06	0.05
	5.7	0.1	0.087	0.08	0.07
Rapeseed	2.2	0.05	0.046	0.042	0.036
	2.7	0.056	0.05	0.046	0.04

meal stored at 20°C could have exploded already at a 0.02% hexane content (Randant, 1982).

Apart from the explosion danger caused only by dust or only by gas, an explosion of a dust/gas/air mixture is also possible. It follows from the investigations on explosiveness of dust/hexane/air mixtures that dust/air mixtures and hexane/air mixtures outside their respective explosion limits can together form a hybrid explosive mixture.

4.1.3 Fumes from fat refining

When released directly to the atmosphere, the fumes from neutralization and deodorization can be noxious in the surroundings of the plant. The odorous substances are short-chain alcohols, fatty acids, aldehydes and ketones.

Fumes formed during soapstock splitting can be neutralized by passing them first through an air condenser, then through a condenser cooled with well water, and finally—after adding a potassium permanganate or sodium hypochlorite solution—to an ejector sucking the fumes (Steinhauer and Graalmann, 1981). The compounds contained in the added solution oxidize the gaseous compounds and make them harmless. The material of the equipment must be resistant to sulphuric acid vapours, since they can occur in the fumes in a concentration of up to 30 g/m^3. Acid fumes are formed during washing and neutralization of acid water. They should be passed through a scrubber, but this increases the volume of liquid wastes.

Fumes released from the deodorizer to the atmosphere constitute an important problem for environment protection. They contain free fatty acids, aldehydes and other components responsible for the unpleasant smell of edible oils. Elimination of the odour of the fumes from deodorization is difficult, since their composition is not well known and depends on a number of factors, like the kind of oil and parameters of the processes preceding deodorization. The detection threshold of some of these substances is very low, in some cases reaching even 1 part per billion (10^9).

Fumes of the deodorization scum have an intense, unpleasant smell. To remove it, the air from the barometric condenser is passed through a cooling tower. However, this method does not completely eliminate the smell and additionally causes the wooden construction of the tower to be covered with fungi. In order to avoid this, the fumes are first passed through two plate heat exchangers operating alternately. Owing to this, wastes rich in fatty matter do not reach the tower, through which flows only pure water used in turn for washing the fouled out-of-use heat exchanger. Substances accumulating in it are passed to the barometric well, from where they are collected for further processing or sale.

Fumes from ventilation of tanks, from condensers and cooling towers, together with fungi colonies possibly formed there, are eliminated by passing them through scrubbers and incinerating in boilers.

Activated charcoal made from coconut shell has a high surface area, 500–1500 square metres per gram, and is a very efficient means of removing all types of gaseous pollutants from the atmosphere (Montenegro, 1985).

4.1.4 Fumes from fat rendering plants

Noxious odours from waste animal fat rendering plants are released during their transport, storage and rendering. The odorous substances are the volatile products of higher fatty acids and triglycerides decomposition, mainly butyric, hexanoic and octanoic acids. These odours are accompanied by protein decomposition products, mainly ammonia, aldehydes, mercaptans, amines, esters, ketones, skatole and indole.

The intensity of the smell of fumes formed during production of fish meal depends on the manner of drying of the processed material. For instance, the amount of volatile substances formed in a direct fired drier is higher by 100–150 mg/m^3 in terms of butyric acid than in a steam drier (Krause, 1978).

The characteristics of the smell depend to a large extent on the manner of rendering (wet–dry) and on the equipment for fume suppression. The fumes can be eliminated in several ways: by incineration, catalytic oxidation, chemical decomposition, multiple washing, wet biological filter or biological filtering bed. Some of these methods cannot be industrially utilized, since e.g. application of scrubbers would increase the pollution of liquid wastes. The most recent method is the application of biological beds in the form of dry filters. They are filled with compost, whose bed height depends on the required time of gas flow. The microflora contained in the compost absorbs or decomposes odorous substances. It should therefore have good growth conditions, i.e. temperature and humidity. Fibrous peat on fir branches serving as a carrier has been proposed during search of the best filter filling materials. Such a design resulted in a decrease of the pressure necessary to force the fumes through the filter from 19 kPa to

1.9–2.9 kPa. Filters of 800 m^2 area and 1 m height of the filtering layer have been applied in a plant (Pfeiffer, 1982) producing 100,000 m^3 of fumes from waste fat processing per hour. The capacity of the filter was 120 m^3 of gases per 1 $m^2 \cdot h$.

Liebe (1986) gives details of the design of a biological filter-bed of this type, suitable for treating fumes from rendering plants, oilseed processing, slaughterhouses, etc. The filter consists of a moistened bed of fibrous peat mixed with heather. The gas stream to be treated must be humidified to 95–100% relative humidity. The various contaminants are removed with 97.9–99.8% efficiency.

4.2 LIQUID WASTES

It is noted that recently developed technologies aim not only at improving the economic results of production, but also at improving water-waste management. The decreasing resources of industrial water in many countries, as well as gradual degradation of the environment caused by all kinds of wastes, particularly industrial liquid wastes, cause not only great losses for human beings and nature, but also losses for the industry in the form of fines charged by the state or the owners of the sewage system. Hence, proper management of the wastes leaving a factory yields two profits: better utilization of the raw material and avoidance of fines charged for the incorrect composition of the liquid wastes. Since the most modern technologies develop in this direction, it is purposeful to discuss the advances in the particular technologies, in which fatty by-products and wastes are formed.

Compared to other branches of the food industry, the Polish oil industry, requiring 25,300,000 m^3 of water annually (1980), is not a great consumer, since this is only 6.2% of the total amount (Skalski, 1975). In 1970 the additional water consumed by this industry originated 76.5% from surface waters, 15% from ground waters, and 8.5% was drinking water. The amount of liquid wastes disposed of by the oil industry was equal in 1980 to 24,800,000 m^3, which constituted 6.7% of the total amount of liquid wastes produced by the food industry. The total load of organic substances disposed of in 1970 was equal to 728 t O_2 annually, the BOD of the wastes being in the order of 40 mg O_2/dm^3. This was 0.22% of the total load of the food industry.

4.2.1 Origin of liquid wastes

Typical examples of surface water pollution are the following:

1. Washed by rain water to sewage systems:
   —spilled seed, oils and fatty acids,
   —dust,
   —the collected wastes,

—results of failures, e.g. fires, breaking of tanks, leakages.

2. Process waters:
   —oils, fatty acids, water soluble organic and inorganic substances,
   —solvents and chemicals.
3. Meal and seeds.

Liquid wastes from solvent extraction originate mainly from miscella refining and degumming. The latter contain 417 mg/kg of solid substances, the number comprising 203 mg/kg of organic substances and 127 mg/kg of suspended matter. Their BOD is 67, and after chemical purification and centrifuging it is less than 15 (Crauer, 1970). Large amounts of wastes consist of the condensate of steam used in the desolventizer and from miscella distillation. They do not contain hexane, since their temperature is much higher than the hexane b.p. On the other hand, they contain small amount of dust from meal and oil, carried away during solvent removal. These substances affect the BOD and should already be removed during waste treatment in the oil mill.

Solvent extraction requires large amounts of water for cooling hexane-containing vapours. This water usually circulates in a closed system, comprising a cooling tower. However, the excess of water drained from the cooling tower can pollute the environment, since this water contains antifouling agents and dust accumulating in the tremendous amounts of air used for cooling the water.

Depending on the processes used, the wastes from solvent extraction can be polluted with other substances, e.g. from wet cleaning of dusty air, from degumming or from washing the production area and the equipment.

Heavy rains, washing the meal or seed dust from the area of the plant, should also be taken into consideration. First portions of water from such rains are usually directed to the sewage treatment plant, while the next are sent directly to a river or a lake. Some of the plants collect all the rain water and then gradually pass it through their sewage treatment plants.

Wastes from solvent extraction are turbid. When they come from rapeseed processing, they have not only the smell of petrol, but also of hydrogen sulphide. Their BOD is equal to 16,000 mg O_2/dm^3, while the fat content is 460–800 mg/dm^3. These wastes undergo a rapid decay.

The greatest problems are connected with wastes formed during refining of fats. Less deleterious are wastes from cleaning the production halls, scrubbers, and from the possible leakages from glands. The least dangerous are the wastes from plate heat exchangers. It can be generally stated that due to a great diversification of the equipment and processes used there are great differences between refineries both with respect to the amount and the composition of the wastes. Moreover, wastes from refineries often change qualitatively due to batch soapstock splitting and possibly to deoiling of bleaching earth. Modern ways of carrying out these processes eliminate such changes.

A refinery produces wastes of a temp. of 40°C, turbid, milky-yellow, smelling like spoilt fat. Their BOD is equal on average to 360 mg O_2/dm^3, fat content to ca. 200–2625 mg/dm^3, while sulphate content — ca. 80 mg/dm^3. Wastes from margarine producting plant have BOD equal to ca. 78 mg O_2/dm^3 and fat content ranging from 40 to 1286 mg/dm^3. The amount of sulphates is ca. 108 mg/dm^3, of suspended matter — 1415 mg/dm^3, while of chlorides — 66 mg/dm^3. Wastes from hydrogenation have a temperature of ca. 40°C, contain CO_2 and no oxygen. Generally, edible fat industry wastes have an almost neutral reaction (pH 6.7), BOD of the order of 334–520 mg O_2/dm^3, and a fat content from 255 to 396 mg/dm^3. Hence, two features are characteristic for these wastes, i.e. high BOD and large fat content. According to another source the BOD is equal to 350–730, and the fat content to 385–499 mg/dm^3 (Bystram, 1976).

Big oil storage tanks in fat processing plants potentially endanger the aquatic fauna and water resources in the case of an uncontrolled oil leak. To avoid such an accident it is necessary to permanently control the tanks and the fittings, so that it is possible to find a leakage as soon as possible. Dikes surrounding the tanks prevent a rapid discharge of large amounts of oil (Boyer, 1980). A similar hazard can be created by oil or solvent spilled during reloading. For this reason the loading areas are surrounded with dikes. The spilled liquids are transferred to skimming tanks. After phase separation the lower layer is discharged to the sewage system, while the upper is further processed. A part of the wastes originates from washing the equipment and cleaning the factory area.

Large amounts of water supplying cooling towers and heat exchangers create a danger of thermal pollution.

The possibility of a heavy contamination of waters by the oil industry has promoted numerous investigations in this area. Since a solvent extraction plant is often combined with a refinery, the quantitative data and the characteristics of wastes from such complexes have been determined in USA (Goodrich, 1980). It follows from these data given in Table 4.2 that wastes from solvent extraction are much less dangerous than from refining, since the former have the BOD on the order of 113

Table 4.2 Wastes formed during solvent extraction of soyabean and refining of oil (Goodrich, 1980)

	Extraction of seed		Refining of oil		Total	
Daily throughput (t/d) * of seed, ** of oil	* 2000		** 350			
Liquid wastes (m^3/d)	330		860		1190	
Characteristics of crude wastes	mg/l	kg/d	mg/l	kg/d	mg/l	kg/d
BOD	340	113	6400	5500	4950	5613
COD	815	267	15000	13000	11400	13267
Suspended matter	210	69.0	3100	2680	2380	2749
Fat	280	92.5	1500	1300	1200	1392

kg/d, while the latter—5500 kg/d in the case of a factory processing 2000 t/d soyabean and 350 t/d oil. In both the cases the determination has been carried out after gravity separation of oil and floating substances.

4.2.2 Waste treatment

As the regulations concerning the allowable level of waste pollution become more and more severe, the methods of waste treatment are continuosly improved.

A general guideline of a rational waste management should be the protection of water quality and its repeated utilization, which enables a minimization of the amount of wastes discharged outside. Some of the methods of achieving this goal are the following:

—application of cooling towers with separation of cooling and polluted water,
—repeated utilization of water from cooling towers (after passing it through a separator) for rinsing and continuous washing,
—recycling of condensate and hot water from heat exchangers for supplementing the boiler water,
—utilization of water from oil processing to soapstock splitting,
—application of measuring devices for the control of water consumption,
—elimination of old tubing, enabling uncontrolled water uptake,
—switching off the ventilators in winter, which reduces losses by evaporation,
—thorough separation of clean and polluted water, which reduces the amount of concentrated wastes,
—reduction of the amount of rain waters passing to concentrated wastes,
—repeated utilization of clean water,
—recovery of fats and wastes from the aqueous phase,
—removal of impurities and useless wastes.

Therefore fat processing plants have the following tasks in the field of environment protection:

1. Solvent extraction plant:
   —clean the raw material and remove the impurities,
   —collect the dust from the air leaving the plant,
   —remove the solvent from the air leaving the plant,
   —recover the solvent and the oil from the liquid wastes being discharged.
2. Refinery and hydrogenation plant:
   —control the pH and BOD of the liquid wastes,
   —remove spent bleaching earth and catalyst,
   —limit oil spills,
   —reduce noxious odours.
3. Fatty acids production plant:
   —control the pH and BOD of the liquid wastes,

—reduce noxious odours.
—recover the fats.

Liquid wastes discharged from the fat industry have usually the following features deleterious for sewage systems: temperature above 30°C, pH outside the range 6.5–8.5, and the presence of detrimental substances in the form of suspensions, solutions and sediments. Oils and fats contained in industrial wastes decompose releasing fatty acids, which attack concrete. Larger amounts of fats can plug the ducts and hinder the biological treatment of the wastes.

The kind of equipment used for waste treatment depends on the requirements of the institutions responsible for the sewage systems or of natural environment protection agencies. Such requirements usually comprise limiting temperature ranges, pH, BOD, sulphate and toxic substances contents.

In the USA and in Western Europe the most often encountered requirements concerning wastes discharged to rivers are the following:

pH	6.5–8.5
dissolved oxygen	> 4 mg/l
temperature	< 30°C
suspended matter	< 30 mg/kg more than in the intake
BOD	< 30 mg/l
COD	< 60 mg/l

The most economical solutions are taken into account when making decisions on the kind and scope of waste treatment. The most important aspects are: operating costs, recent and probably future municipal charges, the value of the recovered fat, reduction of taxes, administration costs, capital costs, etc.

Depending on the degree of pollution and environment protection requirements the wastes can be treated at three stages: pretreatment, biological treatment and possibly tertiary treatment. At each of these stages there are various methods of removal of the particular components of wastes. For instance, pretreatment can comprise: cleaning on grates, neutralization, sedimentation, degreasing, flotation with flocculation, averaging of the composition, coagulation. Biological treatment can be car-

Table 4.3 Characteristics of municipal wastes after the particular treatment stages (Eckenfelder *et al.*, 1970)

Components, properties	Crude wastes	Outflow from pretreatment	Outflow from activated sludge
		mg/l	
BOD	200	132	12
COD	550	275	100
Suspended matter	250	103	< 20
Dissolved substances	620	620	435

ried out using activated sludge, sewage ponds or biofilters. Tertiary treatment comprises adsorption on activated carbon, filtration on sand filters, distillation, ion exchange, reverse osmosis, electrolysis, etc. The characteristics of municipal wastes after two stages of treatment are listed in Table 4.3.

Methods of removal of the particular pollutants from oil industry liquid wastes are the following (Lacy *et al.*, 1975):

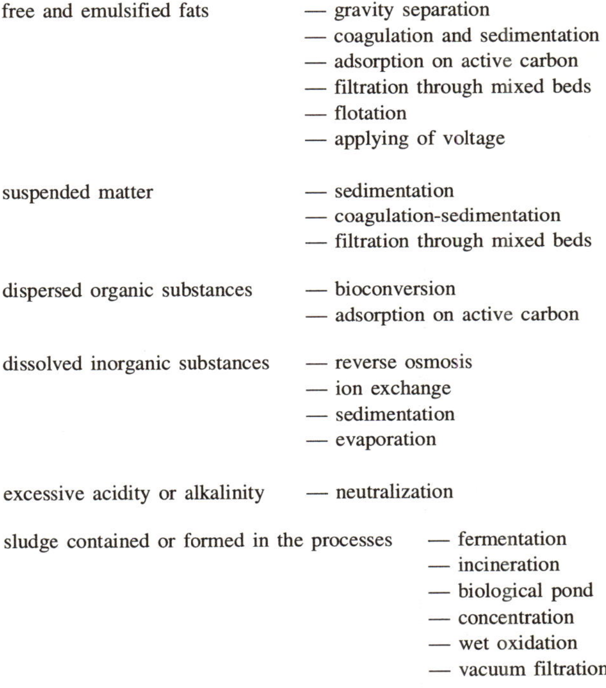

free and emulsified fats	— gravity separation — coagulation and sedimentation — adsorption on active carbon — filtration through mixed beds — flotation — applying of voltage
suspended matter	— sedimentation — coagulation-sedimentation — filtration through mixed beds
dispersed organic substances	— bioconversion — adsorption on active carbon
dissolved inorganic substances	— reverse osmosis — ion exchange — sedimentation — evaporation
excessive acidity or alkalinity	— neutralization
sludge contained or formed in the processes	— fermentation — incineration — biological pond — concentration — wet oxidation — vacuum filtration

Table 4.4 Efficiency of the particular processes of waste treatment (Cantrell and Keller, 1975)

Treatment process	Characteristics of wastes after treatment		
	BOD	suspended matter	fatty matter
		mg/kg	
Crude wastes	2635	1400	485
Biological pond	475	580	105
Filter	296	602	75
Final clarifier	125	110	35
Chlorination	60	90	15
Completeness of treatment (%)	97	94	97

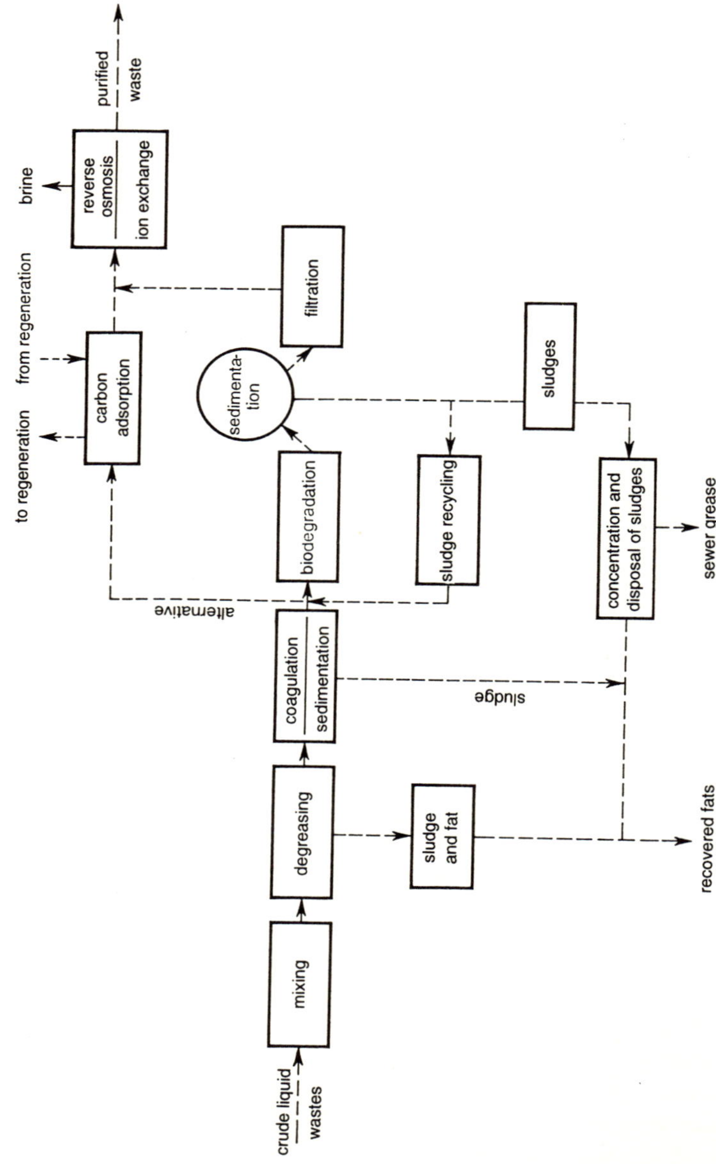

Fig. 4.1 An example of a system of complete purification of edible oil industry wastes (Lacy *et al.*, 1975).

A comparison of the efficiency of these methods is given in Table 4.4, while an example of an outline of the processes of complete waste treatment is in Fig. 4.1.

The choice of the manner of waste treatment depends on numerous local conditions and cannot be standardized. However, in any case it should be kept in mind that wastes discharged from factories are a valuable raw material, either as an utilizable raw material, or as water that can be used again. It can be anticipated that physical methods of purification, like evaporation, filtration, etc., alone or combined with other methods, will allow a complete re-utilization of water and its components, so that it will only be necessary to slightly supplement its amount.

It is recommended that the wastes from solvent extraction are treated by: gravity separation of fat and floating substances, activated sludge and sedimentation pond with a device for the removal of the excess of sludge. Such a treatment allows reducing the BOD to 40 mg/dm^3 and suspended matter to 50 mg/dm^3; however, the regulations in many countries do not accept such levels and the treatment must be more thorough.

The following possibilities should be taken into consideration and calculated when making decisions on the manner of treatment of wastes from edible fat refineries (Boyer and Fithian, 1980):

—discharge of wastes to municipal waters after partial treatment comprising mixing, gravity separation of fat and pH control,
—installation of equipment for biological treatment, assuring purity sufficient for the wastes to be discharged to municipal sewage systems, open waters or sanitary wastes.

Production installations partly purify the wastes from degumming, neutralization and deodorization, by removal of suspended and fatty matter. Wastes from different unit operations should not be mixed before pretreatment. Wastes from hydration can constitute a good example —when mixed with other oil-containing wastes they cause a formation of very stable emulsions.

Neutralization of wastes consists in reducing their amount by application of closed systems and in purification by mechanical, chemical, physicochemical and biological methods.

4.2.2.1 Closed systems

Reduction of the amount of wastes is connected with savings in water consumption. Therefore a development in this field is significant for both the industry and the natural environment.

Oil mills utilizing old technologies with open systems, processing annually 300,000–400,000 t seed and 120,000–150,000 t crude oil, produce ca. 8 million m^3 wastes, 6 million of which come from deodorization. Hence, it is this unit process that creates the possibility of the greatest reduction of water consumption, thus also the amount of wastes (Kroll, 1980). Water consumption of a refinery varies within a broad range, viz. 25–74 m^3/t of the processed oil (Lüde, 1957; Bystram, 1976). According to other sources, water consumption in m^3/t of the processed raw material in open systems is the following (Team work, 1969):

pressing and solvent extraction	40–50
refining	37–74
hydrogenation	7–8
production of margarine	9–22

According to Polish sources, the amount of liquid wastes in a factory processing 400 t seed and producing 200 t margarine per day is equal to 500 m^3/h, 90 m^3/h being technological waste, 400 m^3/h—cooling waters, and the rest—sanitary wastes (Bystram, 1976).

Average water consumption per 1 t of the processed seed is equal in Poland to 54.9 m^3, while the amount of wastes — 52.4 m^3 (Kubicki, 1965). These wastes consist of 69% of cooling waters, 29.2% of technological wastes, 1.3% of sanitary wastes and 0.5% of unidentified wastes.

Assuming that 1 t of soapstock is formed on average during refining of 2 t of oil, it is necessary to extract from it 1%, i.e. 20 kg FFA, and 10 kg of neutral oil, the total being 30 kg. Splitting of 1 t of soapstock requires ca. 30 l sulphuric acid, 300 l hot water and 100 kg steam. Neutralization of acidic water requires 14 kg $CaCO_3$ and 170 kg water. Thus 34 kg of sludge and 1,630 kg of wastes are formed after neutralization (Lau and Neelsen, 1980).

In factories utilizing recycling the share of wastes from solvent extraction plants and refineries is 21%, while in those where recycling is not used it is as much as 53.6%, hence it is possible to save 32.6% of water in these plants, which amounts to 1617 m^3/day for one factory (Kubicki, 1965). Solvent extraction yields an average of 10 m^3 of wastes per 1 t of the raw material in open systems, while 2 m^3 in closed systems; in the case of refineries the figures are 25.6 m^3 and 15 m^3, respectively. Purification of wash waters formed during washing of oil after its alkali neutralization is difficult and costly. On one hand, ca. 0.5% of oil is lost in these wastes, and on the other — they largely increase the BOD. A system of recycling these waters has been proposed in order to eliminate these wastes (Eisenhauer *et al.*, 1970). After washing, the water is passed through a cation-exchanger bed absorbing sodium, while the slightly acidic water is recycled. A reduction of sodium content from 34 to 1.5 mg/kg has been achieved during experiments with soyabean oil. This result is comparable or better than that obtained when washing the oil with fresh water. The system has been developed on an industrial scale.

A circulation system presented in Fig. 4.2 has also been applied. Wash waters, separated in a separator, are pumped through two columns operated alternately, packed with an ion exchange bed. The bed absorbs soap cations, 93.7% of the sodium being removed at pH 3. The iron content falls to

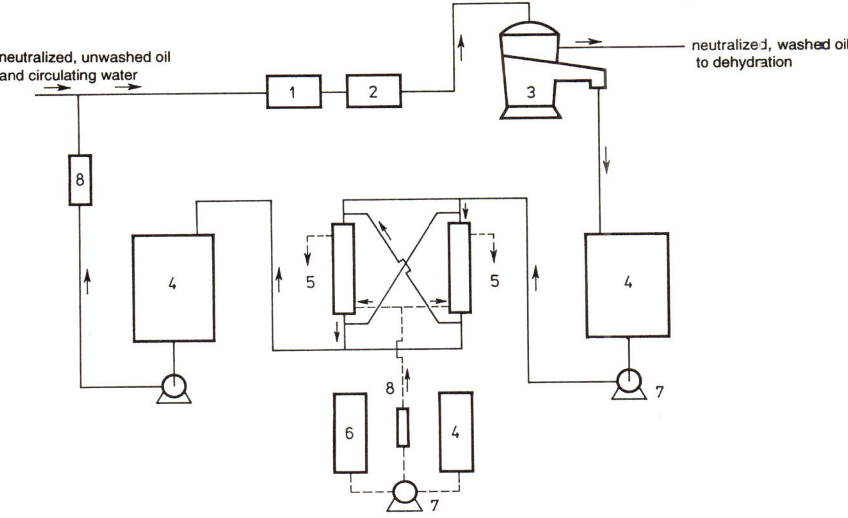

Fig. 4.2 Recycle washing flowsheet (Beal *et al.*, 1973): 1—heater, 2—stirrer, 3—centrifuge, 4—water tank, 5—cation exchange resin column, 6—acid regeneration apparatus, 7—pump, 8—flow meter.

0.02–0.09 mg/kg, and copper content to 0.02 mg/kg. Oil refined in such a way behaves the same as traditionally refined oil during hydrogenation. Soluble substances accumulate in the water when washing in this manner. The analysis of such water revealed that they are mainly phospholipids; no fats have been found. Column operation time was equal to 8–9 h at 7000 kg/h oil flow. The BOD is reduced by this method by 94% (Beal *et al.*, 1973).

In continuos soapstock processing by the de Laval method a part of the acidic washings is recycled to the main soapstock stream, which reduces the sulphuric acid consumption by ca. 10–15%. Since acidic water from non-degummed oil has a high BOD (Table 4.5), it is treated separately by neutralization with caustic soda to pH 7, and only after that it is introduced to the main stream of wastes. The amount of dry NaOH consumed is usually 0.3–0.5% of the mass of this waste. A schematic diagram of processing of

Table 4.5 Characteristics of acid water from splitting of non-degummed soyabean oil (Crauer, 1970)

Components, properties	Content (%)
Solids	4.0–12.3
Organic substances	1.4–6.2
Suspended matter	0.2–0.9
Fat	0.1–0.9
Phosphorus	0.11–0.43
Invert sugar	0.04–0.64
BOD (mg/kg)	16.400–52.400

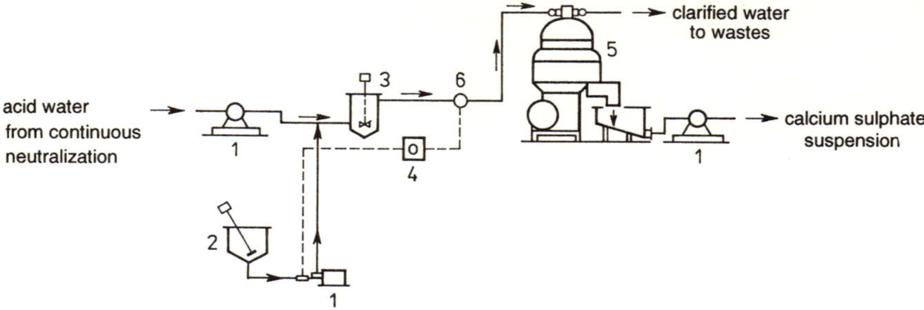

Fig. 4.3 Treatment of acid water from soapstock hydrolysis (Crauer, 1970): 1—pump, 2—lime suspension tank, 3—stirrer, 4—pH adjustment, 5—clarifier centrifuge, 6—pH meter.

acidic water from oil degummed by the continuous de Laval method is presented in Fig. 4.3.

Neutralization is carried out using slaked lime, the addition necessary to reach pH 7.0–7.6 being controlled automatically. Thorough mixing of the lime with the acidic water takes place in a mixer, from where the mixture passes to a clarifying separator. The aqueous phase is passed from this separator to wastes (sometimes through a heat exchanger), while the insoluble sediment is periodically removed. It contains dibasic calcium phosphate, which can be utilized as a fodder additive or as a fertilizer. The amount of this sediment varies within a wide range from 3 to 10% v/v. The temperature of the process is 50–80°C, and the amount of lime introduced is 0.6–1.4% of the mass of the acidic water. The method described for processing of acidic water has the following advantages:

—it reduces the BOD by 62–76%,
—it removes 80–95% of the invert sugar,
—it removes free phospholipid bonds,
—it yields a valuable by-product,
—it neutralizes water and reduces the content of insoluble substances to a level below 300 mg/kg.

Recycling of water from deodorizer barometric wells allows a significant reduction of the amount of wastes. The water can be used as (Kroll, 1980):

—cooling water — by recycling and cooling in a cooling tower the uptake of this water can be reduced to losses due to evaporation, spraying and flushing of sludges;
—water for barometric condensers—in three ways:

1. *Single circulation.* Liquid wastes from a barometric well are passed through a degreaser, cooled, and directed to spraying the condenser. Noxious odours, spraying of lipids contained in the condensate in the vicinity of the cooling tower and increasing the load of the tower by

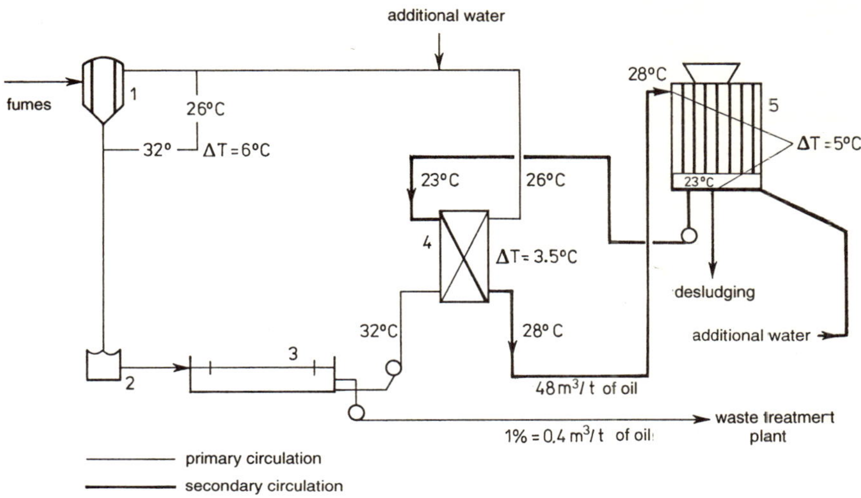

Fig. 4.4 Cooling and recirculation of deodorization eluate (Kroll, 1980): 1—barometric condenser, 2—condenser water tank, 3—degreaser, 4—plate heat exchanger, 5—cooling tower.

delivering lipids to the separators are the drawbacks of such a solution. Besides, waters from desludging of the tower must be treated.

2. *Double circulation.* The above disadvantages of the single circulation system can be eliminated in a double circulation system. Waters from wells are cooled in plate heat exchangers, and used again for spraying in barometric condensers. Water from plate condenser is cooled down in a cooling tower and recycled. Lipids gradually accumulate in the plate condenser, therefore it must be periodically washed with water, soda lye and hot water. Since the amount of water in the first circulation increases gradually by the amount of steam condensate from deodorization, the excess of condensate is passed to a waste treatment plant. Figure 4.4 presents the circulation of water in the double system.
3. *Japanese "Gasoline" system.* It is a variant of the double circulation system. NaOH is added to the deodorizer scum. As a result, fatty acids contained in the scum change into soaps. These soaps absorb the lipids, forming with them a uniform, stable emulsion. Owing to this, the degreaser and the additional plate heat exchanger become unnecessary. However, in this case also it is necessary to drain a part of the increased volume of the condensate to a waste treatment plant. The consumption of water supplying the condenser, measured as the ratio of "acid" water to water in the barometric condenser, can be reduced from 1:10 to 1:1–1:3.

Application of recycling allows saving 19.5% water consumed by batch refineries. In the case of solvent extraction this saving is 12.7%.

4.2.2.2 Mechanical cleaning

To protect the pumps, pipelines, fittings and stirrers against damage due to the presence of coarse impurities, the wastes are first passed through gratings and sieves. The distance between the rods of the gratings is 8–15 mm. The gratings are equipped with devices for cleaning them, similar to haircombs. A sand separator is used to separate fine impurities of higher density, e.g. sand particles (Brauch, 1982). It is designed in such a way that it stops 99% of particles of the diameter of 200 μm and more, which is equal in terms of the standards for municipal sewage systems to 95% of particles of the diameter of 150 μm. A partial removal of the floating fat layer takes place in the sand separator. This fat is then passed to a special tank.

Removal of fatty matter. Fat can occur in the aqueous phase in three forms:

—as free fat, forming droplets of the diameter > 5 μm, accumulating at the surface in the fat separator and gathered from there,
—as droplets of the diameter < 5 μm, not moving upwards, but remaining in the suspended form. They undergo agglomeration e.g. under the influence of high-speed rotary pumps,
—as an emulsion, stabilized e.g. by surfactants. In this form it is impossible to separate the phase using separators. Separation is achieved e.g. using biological methods or adsorption on active carbon.

Removal of fatty matter can be accomplished in two stages, by gravity separation and flocculation with flotation. The first of these methods can be treated as a mechanical method, while the second — as a physico-chemical method.

Gravity separation utilizes the difference in densities of fat (0.915) and water, due to which fat rises to the surface. The separation rate, hence the volume of the equipment, depends on the amount of the oil phase, pH and the characteristics of water. It is usually a few tens of seconds. On the other hand, separation does not take place when the water contains clay, since the settling deposit binds oil.

All the wastes from a refinery are first passed to a clarifier, whose volume should ensure a 2–3 h residence time. Since the difference between the densities of water and oil is small, the gravity separation is slow,

Table 4.6 Typical results of impurities removal by skimming (Cantrell and Keller, 1975)

Impurity	Content (mg/kg)		% removal
	inlet	outlet	
BOD	4010	2440	38
Suspended matter	3680	2700	33
Fatty matter	3985	3200	25

ca. 2–3 m/h. The separation rate is the greatest at pH 2–3 and at a temperature of 50–60°C (Wall, 1982). A reduction of the ether extract down to 25 mg/l of waste is considered a sufficient degreasing degree (Lüde, 1957). A comparison of the results of analysis of wastes treated by various methods is a good proof of the development in this field. Table 4.6 lists typical results obtained by skimming. According to Wall (1982) the degree of fat extraction is 60–90%.

Separation methods can differ depending on the conditions (Choffel, 1976):

—A degreaser with mechanical collection of the fat is used when the time of rising of the fat to the surface is shorter than one minute. The dimensions of the chamber are 0.3 $m^2/m^3 \cdot h$. The wastes are passed through at least three compartments.
—The same device, but with an addition of flocculants and with injection of compressed air is used when the separation is slow.
—Clean water with recirculation and cooling should be used when the separation does not occur and the oil settles at the bottom.

Gravity separation can be rationalized in modern systems for example as follows: the wastes are discharged to a concrete tank containing 28–45 corrugated sheets inclined at an angle of 45°. The fatty phase migrates along the bottom surface of the sheets to the surface, where it is gathered, while the aqueous phase and the sludges accumulate at the bottom of the tank, from where they are disposed.

Filtration. Some factories pass their wastes from aerated ponds to filtration before discharging to open waters. A bed made of a granular material is used as the filter, the particle size being intermediate between that of sand and 3 mm. The bed is ca. 80 cm high, and consists of a layer of the granular material. Usually very fine sand is situated at the bottom, then normal sand and finally anthracite. The wastes pass downwards. In such a system coarse impurities are stopped in the upper layers, and fine impurities close to the bottom. Large amounts of wastes can be filtered under such conditions before the filter becomes clogged by the impurities, mainly microorganisms. A small amount of a coagulating agent is added to the wastes prior to filtration in order to accelerate this operation. Suspended matter is not always efficiently stopped by such a filter. Sometimes 50% is stopped, sometimes less.

Filter presses with countercurrent cleaning of their packing were applied in search of better methods of purification (Cantrell and Keller, 1975).

Microfiltration or ultrafiltration have been proposed for the purification of industrial wastes containing oils (Rautenbach *et al.*, 1984). Using this method it is possible to decrease the oil content in wastes from the initial concentration of e.g. 1% to less than 50 mg/dm^3 when using only microfiltration, or to less than 5 mg/dm^3 when utilizing also ultrafiltration.

Rotating scrubber. A reduction of the load of lipids in wastes from deodorizers can be achieved in several ways. One of them is the application of a cooler — rotating scrubber. Such a scrubber is extremely effective. A semi-continuous deodorizer of the capacity 150 t/d, working at a pressure of 0.7 kPa, requires 300 m^3 of water per day. The fat content of the resultant liquid waste is 550 kg/day. When the same apparatus is equipped with a cooler — vapour scrubber, the fat content of the waters in the barometric wells falls to 12–15 mg/kg, which reduces the amount of fat to 108 kg/d, hence five times less.

4.2.2.3 Chemical treatment

Before the chemical or biological treatment the wastes should be neutralized. The pH required when the wastes are discharged to natural waters is 6–9. In biological treatment plants the pH should be kept within the limits 6.5–8.5.

Neutralization. Wastes, particularly those from batch processing in the refinery, have very variable pH. Soap splitting yields wastes of pH equal to 3 or less, while wastes from neutralization are strongly alkaline. Therefore such wastes should be collected in a suitable apparatus in order to standardize their pH and partly neutralize. This results in significant savings of the chemicals, which would have to be used if each of the waste streams were to be treated separately. The wastes can be stirred with mechanical devices or with air, provided it does not cause unfavourable side reactions.

Acidic wastes can be neutralized with lime or burnt magnesite. They can also be neutralized with alkaline waters from cleaning of deodorizers. Alkaline wastes are neutralized with sulphuric acid provided that the sulphate content limit is not exceeded, or with hydrochloric acid. Neutralization should be carried out after clarification, when the fatty emulsions are already broken.

Exhaust gases are used for neutralization in factories having their own boiler houses. Exhaust gases containing carbon dioxide can be used for neutralization of alkaline wastes. Salt is not formed in such a process, so the wastes do not transport it to the following parts of the waste treatment plant. Sulphur dioxide is additionally removed from the exhaust gases, due to which no sulphuric acid, responsible for large corrosion losses, is formed.

Equipment made of a plastic resistant to 50% NaOH is used for neutralization. The alkali demand is determined by an automatic pH-meter. The wastes should be degreased to protect the electrodes against fouling. Separation proceeds better at low pH; however, it must then be raised to a value required by the standards. Some plant is equipped with automatic valves, letting the wastes out only when they reach the required pH. Acid waste from splitting is neutralized and next the insoluble substances are separated by sedimentation and added to other wastes used for animal feeding. After neutralization and separation of the suspended matter the aqueous phase readily undergoes biodegradation.

When the amount of wastes from deodorization is small, the impact of the impurities contained in the acidic water on the wastes from the entire factory significantly increases. Therefore it becomes necessary to remove the main components of the acidic water, i.e. sulphates and organic substances. Sulphates in wastes are responsible for the corrosion of concrete. The presence of sulphates can be avoided in the following ways:

—replacement of alkali neutralization with distillatory neutralization,
—conversion of sulphates into chlorides by precipitation of calcium sulphate using calcium chloride. The main disadvantage of such a solution is the formation of large amounts of $CaSO_4$, whose separation, drying and transport to a dumping ground are expensive,
—soapstock splitting using hydrochloric acid instead of sulphuric acid. Technically this is a very good solution, but it is uneconomical.

The kind of neutralization method used and the sequence of the particular processes of waste treatment depend on the characteristics of the wastes.

4.2.2.4 Physicochemical treatment

Undissolved fat and higher fatty acids or their salts can be recovered from wastes by skimming. Mechanical separators or flotation based on hydrophobic and aerophilic properties are used for this purpose. Breaking of emulsions or splitting of soaps increases the salt content of wastes and causes the formation of a useless sludge, which is difficult to transport to dumping grounds, and releases noxious gases during incineration.

The very fine suspension readily starts to mix not only when the waste moves, but also under the influence of temperature differences. Chemical coagulation, causing flocculation and flotation or sedimentation, is used

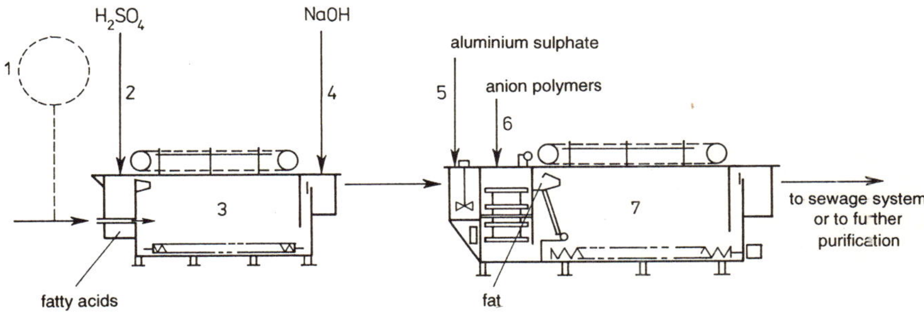

Fig. 4.5 Pretreatment of wastes from vegetable oil refinery (Wall, 1982): 1—equalizing tank, 2—pH adjustment, 3—skimming tank, 4—neutralization, 5—coagulation, 6—flocculation, 7—flotation unit.

to remove such impurities. The coupling of degreasing with coagulation, flocculation and flotation, presented in Fig. 4.5, is quite frequent.

Chemical coagulation and flocculation. Stabilized fat emulsions in the aqueous phase are particularly difficult to break when the system contains surfactants—e.g. phospholipids, detergents used for cleaning, etc. Addition of polyelectrolytes (polymers of the mass of ca. 1 mln) promoting flocculation is very efficient in such cases. The polymer-surfactant complex formed precipitates from the solution, facilitating the formation of floccules (Axberg *et al.*, 1980). The choice of a proper electrolyte (anionic or cationic) depends on the character of the surfactant present in the system. Better results are obtained when using anionic and non-ionic polymers than cationic polymers, especially at low pH.

Table 4.7 Typical results of impurities removal from degreased wastes by flotation with air in the presence of aluminium sulphate and polymers (Cantrell and Keller, 1975)

Impurity	Content (mg/kg)		% removal
	inlet	outlet	
BOD	2440	740	70
Suspended matter	2706	400	84
Fatty matter	3195	360	88

They reduce the turbidity by 10–50% more than the coagulants alone. Addition of aluminium sulphate serving as a coagulant is equal at pH 5.5–6.5 to 10–30 mg/kg. It works particularly well with polyelectrolytes added in an amount of 0.5 mg/kg.

Wastes subjected to flocculation are passed through a vertical sand filter. Table 4.7 lists the results obtained by application of flotation with air and utilization of aluminium sulphate with polymers.

It has been found that wastes from solvent extraction plant can be purified using 90–100 mg/kg coagulating agent, then 1 mg/kg polyelectrolyte and subsequent centrifuging in a separator. The treated waste contains neither fat, nor suspended matter, and its BOD is reduced by 78–90%. The precipitated substances, most often proteins, can be added to the desolventizer-toaster (Crauer, 1970).

Water from barometric condensers usually contains 500–2000 mg/l fatty matter and other substances removed during deodorization. These impurities hinder the outflow of the condensate and make necessary the frequent and expensive cleaning of the equipment. An addition of Amerfloc 10 cationic polymer to water sprayed in the condenser reduces the impurities content from 900 to 10 mg/kg. The fatty matter content is therefore reduced by 99%, while the COD — by 75%. The water changes from turbid, cream-yellow, to completely clear. Fatty matter contained in fumes is readily recovered, and its value exceeds the costs of this purification method (Farr and Bolhofner, 1980).

Zimińska *et al.* (1984) found out that $FeCl_3$ is the best coagulating agent for the fish industry. It is still more efficient when used together with an addition of Rokrysil WF-2 anionic polyelectrolyte. An addition of 4–6 mg/dm^3 with vigorous stirring causes agglomeration of the precipitate and its sedimentation, so that flotation becomes unnecessary. The volume of the sediment is reduced 8 times and constitutes 2–6% of the volume of the waste. On a laboratory scale the COD is reduced by ca. 80% due to a removal of 50% proteins and 80% fat by this method. The sediment separated from the wastes consists of 25.3% fat, 41.5% proteins and 22.6% ash. The waste after coagulation is a transparent liquid of pH 6.2–6.5. It meets the requirements for a waste discharged to a sewage system, except for a much higher chloride content. Sometimes the amount of petroleum ether soluble substances is also excessive. Better results have been obtained on a pilot plant scale, since the ether extract reduced from 1400 to 70 mg/dm^3, i.e. by 95%. Although the paper concerned the fish industry, yet the results can be utilized also in the treatment of wastes from the fat industry — it is only that the removal of fat must be more thorough, which can be achieved by the remaining treatment processes.

Chemical coagulation of wastes is used also after biological treatment. Aluminium hydroxide with an addition of a polyelectrolyte or lime is used as the coagulant (Eckenfelder *et al.*, 1970). The dose of aluminium hydroxide for wastes from oxidation ponds is 100–250 mg/l. It causes flocculation of the matter suspended in the wastes. The coagulants are separated subsequently by flotation or sedimentation.

Flotation. Flocculation and gravity separation combined with it are always accomplished in the presence of air forced through the wastes. After a pretreatment, the wastes can be subjected to flotation, utilizing the hydrophobic properties of fats and the simultaneous tendency to adhere to air bubbles.

Similarly, hydrophobic properties can be given also to other impurities by acting with suitable chemicals (Schmidt-Holthausen, 1979). Flocculation combined either with sedimentation, or with flotation is used for this purpose. In the case of sedimentation the fat particles, which have a natural tendency to rise to the surface, must be weighted so that they accumulate at the bottom. Consumption of the chemicals in such a case is much higher than is necessary for breaking an emulsion or coagulation. On the other hand, the cost of the chemicals necessary for flotation is lower by 20–30% than of that for sedimentation. Morever, the concentration of the floating sludge is higher than that of the sedimented sludge, which is advantageous in the following processes (Richter, 1981). Since the rate of rising of the particles is 3–5 times higher than the sedimentation rate, the area of sludge accumulation can be equal to 25–40% of the area necessary for sedimentation. Hence, in spite of higher costs of the equipment (but much lower building costs), a flotation plant is cheaper than a sedimentation plant.

Flotation is carried out in long, rectangular chambers, supplied with wastes at a depth of 2–2.5 m at the narrower side. A part of the wastes, ca. 15%, is circulated through an aeration vessel, where air is introduced under a small pressure and dissolved. This aerated portion is introduced through a perforated tube into the waste stream. Decompressing air bubbles combine with fat particles and transport them to the surface. A thin fat layer should be left at the surface. The transported fat particles combine with this layer, which prevents turbulences resulting in renewed mixing of the impurities. The layer gathered from the surface is directed to utilization. Sediments accumulate in the chambers. They are gathered with a collector and transported to dumping grounds. In another design the flotation chambers are equipped with sludge catchers. The outflow from flotation contains only dissolved impurities. Their content is determined on the basis of the BOD.

In the case of larger amounts of fat in wastes, they can be preaerated before the initial settling tank (Imhoff, 1982). Such a degreaser of as large as possible an area should assure a flow-through time of ca. 3 min. Compressed air is introduced at the bottom. The fat accumulates in side chambers and is passed to a special tank, while the sediment falls down along the inclined walls of the bottom of the chamber to the central furrow. Air consumption is equal to 0.2 m^3 per 1 m^3 of the wastes.

The temperature of the waste exerts an important influence on the amount of the dissolved air. The lower the temperature, the higher the volume of the dissolved air. On decompression the air bubbles adhere to hydrophobic fat particles and transport them to the surface. A foamy layer of fatty substances is formed in such a way. It can be removed mechanically.

A complete flotation effect is achieved when the air is dispersed into bubbles of 0.1–0.2 mm diameter, which combine also with hydrophilic particles, transporting them upwards. The bubbles can be larger for hydrophilic particles. The air is introduced under a pressure of 0.26–0.41 MPa. A part of the purified waste is recycled to a tank, where it is saturated with air.

The fat — separated gravitationally or by flotation — is dehydrated and introduced to an apparatus equipped with steam coil in order to boil out the residue of water. The fat from gravity separation is better than that from flotation, since aeration adversely affects the fat quality by increasing the extent of oxidation.

A reduction of BOD by 77–97%, of fat content by-100% and of suspended matter by 50–95% has been achieved on a laboratory scale after flotation with an addition of 5% of whitewash, without any other flocculant (Dakovič *et al.*, 1985).

The sludge formed in the various operations must be eliminated. This can be accomplished by various means:

—by recirculating it to production in order to utilize some of its components,

—by digestion, dehydration, drying and incineration or disposal to a dumping ground.

Addition of some chemicals can improve the handling of the sludge by a reduction of its volume, an improvement of the extent of purification, an increase of its concentration, a reduction of the amount of water recycled during flotation and better utilization of the area, which reduces the investment costs. However, chemicals cannot be utilized when the sludge is recycled to processes which they can affect adversely. The same concerns the sludge used as a fodder additive. The sequence of the treatment stages can differ depending on the characteristics of the wastes. For instance, in a Swedish factory flotation is carried out twice, while there is no biological treatment. The fat content is reduced from 500–600 g/m^3 to 6–12 g/m^3, and the recovered fatty phase contains 90–95% fats (Bystram, 1979).

4.2.2.5 Biological treatment

The method is based on the action of aerobic bacteria, which consume the organic carbon contained in the wastes, utilizing it as a source of energy and a substrate for cell synthesis. Aerobic bacteria utilize oxygen for oxidation and synthesis. The bacteria convert organic substances into CO_2, water, ammonium, phosphorus salts and other simple compounds. The microorganisms are introduced to the wastes in a suspended form or bound to a solid support, washed with the wastes supplied with air — either by vigorous stirring, or through bubblers. Numerous flora species take part in the biological treatment. Most of them are heterotrophic. On the other hand, anaerobic organisms derive energy from the decomposition of organic compounds under anaerobic conditions (Eckenfelder *et al.*, 1970).

A "fat" is not precisely defined as a component of wastes. Petroleum ether soluble components are not only proper fats, i.e. glycerides and FFA, but also mineral oils, essential oils, waxes, etc. Therefore a gas chromatographic analysis is necessary when the effect of fats only on the microflora used for biological treatment is to be assessed. Fat-containing wastes are treated according to the scheme presented in Fig. 4.6. It follows from this scheme that fat in the form of soluble fatty acids and sometimes of free fat can be removed only by biological methods. The bacteria contain almost no FFA, and fat contained in them is water insoluble (Krause, 1982).

Under the influence of biologically degradable wastes river water loses its dissolved oxygen, and the growth of bacteria intensifies. Depending on the ratio of the amount of wastes to the amount of river water, the BOD returns to the value preceding the waste discharge only after a few days or after flowing a few kilometres (Fig. 4.7). Biological treatment accelerates the process that takes place naturally in the river. Either intensive biodegradation by highly activated sludge, or — in the simplest case — biological ponds, can be used for this purpose. The basic difference between the application of activated sludge and the pond consists in the fact that the bacteria are not recycled in the pond, and the residence time in

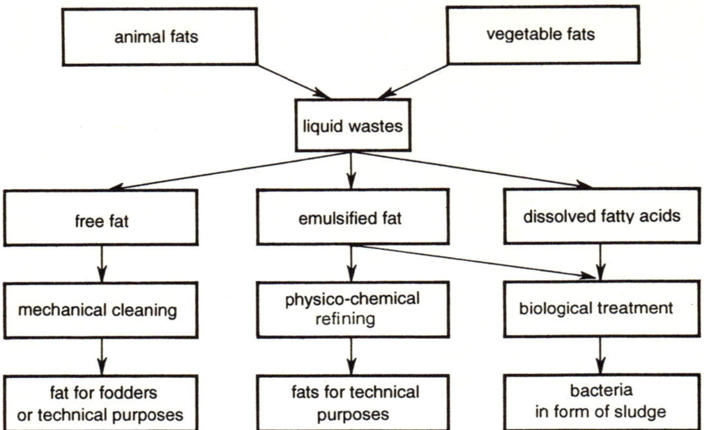

Fig. 4.6 Treatment of wastes containing fats in various forms (Krause, 1982).

the pond is usually 10–20 days. In a treatment plant with activated sludge the flow-through time is equal to a fraction of twenty four hours, and the amount of bacteria is ca. 5,000 mg/dm^3. The production of bacteria is 25% with respect to the microflora permanently staying in the plant. The growth of bacteria requires nitrogen and phosphorus. The nitrogen content is supplemented using e.g. ammonia. The demand is 1 mg phosphorus/dm^3 and 10–20 mg nitrogen/dm^3 per 100 mg/dm^3 BOD. The BOD resulting mainly from the presence of fats is reduced in this process by 95%. Large amounts of bacteria are produced during biological treatment. Together with the suspended matter they form a sludge. In some factories this sludge is anaerobically fermented, which results in a partial decomposition. The residue can be more readily dehydrated. The organic substances formed in this process are converted mainly into methane. The residue is relatively hard and can be utilized e.g. for fertilization of soil.

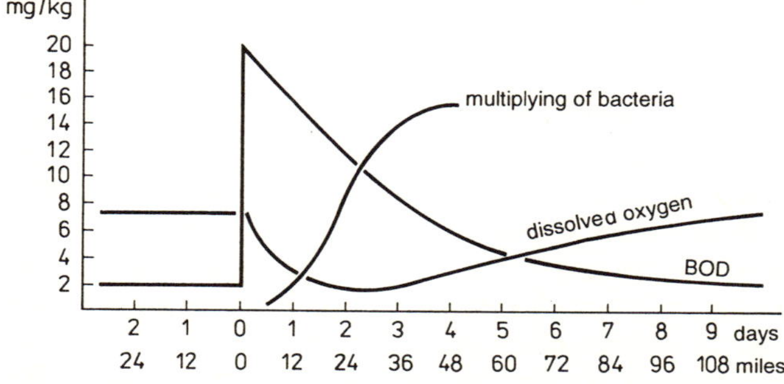

Fig. 4.7 Growth of microflora, BOD and the amount of the dissolved oxygen vs. time and the distance of water flow from the waste disposal site (Krause, 1982).

No large area is required when using activated sludge. On the other hand, when there is no problem with the area and noxious odours, a biological pond also yields good results, decreasing the BOD by 90%. However, it sometimes requires hundreds of hectares of land. Application of aerated ponds is an intermediate solution.

Activated sludge. Utilization of activated sludge requires large tanks, in which the wastes can stay for 4–20 h with vigorous aeration. Under such conditions the microflora playing the basic role grows very rapidly and forms the activated sludge. The wastes flow through an aeration system to a secondary settling tank, where they undergo sedimentation. The trapped sludge is recycled from this tank to the aerated bed in order to increase its biomass. The approximate concentration of the biomass is 3–4 g/dm^3. The most important parameter determining the capacity of a biological treatment plant is the load of the activated sludge expressed as the BOD per 1 kg d.w. per day. This quantity

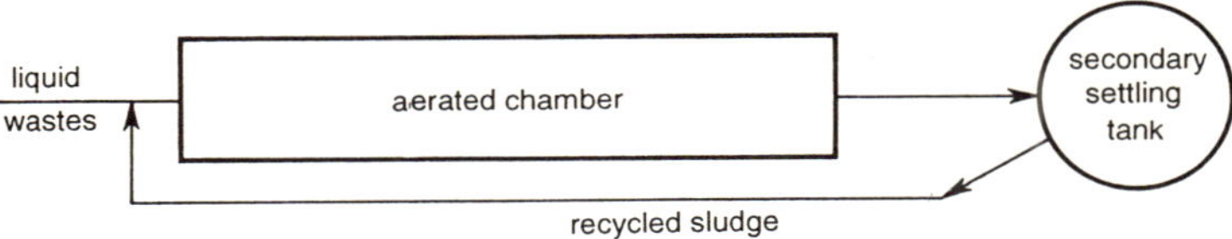

Fig. 4.8 Activated sludge system (Eckenfelder *et al.*, 1970).

determines how much organic matter can be processed by the microflora in 1 m^3 of the chamber during twenty four hours.

The action of the activated sludge system consists in a continuous aeration of the flocculated microorganisms after mixing them with the wastes. There are numerous modifications of this process. In the traditional way a contact stabilization and a prolonged aeration are used. This method is the most widespread in fat industry. The process is carried out in long, rectangular, aerated chambers (Fig. 4.8). Round chambers of a diameter reaching 50 m, as well as square chambers, are also used. The oxygen consumption rate is initially high, but it decreases with time. The system allows reaching a BOD of 20 mg/l in the outflowing wastes.

The amount of oxygen supplied exerts a great influence on the efficiency of the activated sludge. It can be measured by the consumption of the electric energy used for driving the blowers. The blowers supply oxygen under the surface with an efficiency of 2.5–2.7 kg O_2/kW·h. Compared to surface aeration, this method has the advantage of not forming an aerosol (i.e. no waste is dispersed to the atmosphere) and noise. Pure oxygen can be used instead of air, which results in a reduction of the amount of gases passing to the atmosphere. This is very advantageous, since the wastes produce noxious odours. A so-called Bioxon system, capable of automatically controlling the uniform oxygen supply, is used in order to maintain an optimal oxygen concentration in the entire chamber.

The wastes stay in the chamber with activated sludge for a few hours. Continuous mixing accelerates the process. If two or more chambers are used in series, the sludge is partly separated and recycled at each passage.

It has been established that emulsified and sometimes dissolved fat can be decomposed in municipal biological treatment plants with activated sludge, provided that they have a sufficient throughput and that the separable fat is first thoroughly removed from the wastes. In this case the greatest role is played not by the total amount of the delivered, emulsified fat, but by the rate of its inflow to the biological treatment plant per the bacterial mass, hence the amount inflowing in a unit time to the plant (Krause, 1982).

Anaerobic treatment of effluents. Effluents containing a high organic load are common in most branches of the food industry. Lepke (1988) describes an anaerobic system of treatment which is suitable among others for margarine and fat processing plants and for meat processing plants. The process depends on two groups of bacteria which have different optimum growth requirements, and is therefore best carried out in two stages. The first stage of hydrolysis and acidification uses facultative anaerobes which can tolerate pH 3–7. Typical conditions used are 35°C and pH 4.5–5.5. The second stage uses acetogenic and methanogenic bacteria which are obligative anaerobes. The bacteria are thermophilic and operate best at 55°C and pH 6.8–7.2. The process has two advantages in that compared with an aerobic process it produces only one-fifth the amount of sludge for disposal, and the methane generated can be recovered and used for electrical or steam generation.

Biological ponds. Aerated ponds are often used for biological treatment. This requires suitable grounds in the vicinity of the factory. The ponds are usually 4 m deep and are equipped with aerating devices. The efficiency of such a device with respect to the depth of the pond and the demand of oxygen necessary to oxidize the organic substances contained in the waste are important in this case (Kenyon, 1970).

The concentration of bacteria reaches 100 mg/dm^3 in the pond. The purification rate measured in BOD units is equal to 0.2 kg per 1 day and 1 kg of bacteria. After passing through a tranquillization zone, and sometimes through a separate settling tank, where solid impurities are sedimented, the wastes contain 30–50 mg/dm^3 suspended matter and have a BOD of ca. 40 mg/dm^3 (Mc Dermott, 1976).

In another design the wastes from an aerated pond, where they remain for 10 days, are passed to a chamber with activated sludge for another 40 h. Next they pass through a clarifier, then they are chlorinated, pass to a contact chamber and then to a factory pond, from where they are finally discharged to a river. This system can reduce the BOD from 4960 mg/dm^3 to 7.5 mg/dm^3.

Surface aerators are the most suitable devices in the case of activated sludge and biological ponds. In the majority of designs oxygen is transferred in two ways, viz. in droplets and in the thin layers formed as a result of the action of the propeller blades and on the turbulent surface of liquid from air bubbles, entrained in the aerating chamber. Aerators vigorously stir the wastes throwing them up in the form of droplets. They also cause deep stirring. They can be mounted on solid supports or on floating elements. One such device consumes 3–4 kW·h and is sufficient for the aeration of an area of 6 m diameter.

Sprinkled bed. It consists of a coarse packing (e.g. coke, gravel, etc.), contained in tanks and sprinkled with the waste. A biological film is formed on the surface of the bed. The waste flowing through is freed from organic matter, oxidized by the bacteria. The features of such a system are low energy costs at high investment costs, lower operational reliability and worse protection against noxious odours.

Elements made of rigid PVC are also used as a packing for sprinkled beds. Their usable area is 200 m^2/m^3, void volume — 94–98%, diameter — 0.50–1.50 m, and height — 0.65 and 1.35 m (Schoning and Neumann, 1983). Combined with activated sludge they can completely purify the wastes. The first element of such a system decreases the BOD by 60–80%, consuming 0.20 kW·h per kg of BOD.

Biodegradation can be carried out in a cooling tower by passing through it the wastes diluted with water of a temperature 30–40°C, with an addition of phosphoric acid and ammonia (0.4 mg/kg phosphorus and 6.0 mg/kg nitrogen relative to the circulating water). Under such conditions the biodegradation is very rapid — the BOD is reduced by 95%, while the COD by 92.1%. This system yields water savings and eliminates the necessity of wastes treatment and disposal (Holt and Brown, 1971; Neuner and Holt, 1975).

Organic matter can also be removed by other, non-biological methods, viz, using adsorbents of a highly developed surface. However, these methods are very expensive.

The stages of a complete treatment of wastes from an oil refinery can be, for example, the following (Mc Dermott, 1976):

1. Initial pH adjustment	— flow in the apparatus 8.2 l/s — adjustment of pH within the limits 1.5–3.0 to facilitate the gravity separation of fat and water
2. Averaging of the composition	— tank of 850 m^3 volume
3. Degreasing	— apparatus of 1135 m^3 volume with a fixed overfall level for a continuous degreasing with steam and gravity separation — tank for fat collecting of 38 m^3 volume — water recirculated to the second stage

4. Secondary pH adjustment	— automatically controlled introduction of ammonia in order to increase the pH to 7
5. Flotation in two apparatuses with an introduction of air and chemicals	— apparatus of 68 m^3 volume — flocculation by an addition of CaO and alum — coagulation by means of polyelectrolytes
6. Biodegradation in two aerated ponds connected in series	— ponds of 4542 m^3 volume with 5 aerators floating on the surface — purification time — 5–6 days in each pond
7. Stabilization in a pond	— conditions same as in the 6th stage, but without aeration — total residence time in the three ponds — 15–18 days
8. Filtration with chlorination before and after the process	— removal of suspended matter and bacteria
9. Final characteristics of the waste	

	mg/dm^3
suspended matter	50
fat	1.0
phosphorus	9
nickel	0.02
BOD	40
pH	7–8

4.2.2.6 Biological treatment of palm oil mill effluent

Because the extraction of palm oil takes place in an aqueous system, a particularly heavy load of aqueous waste has to be dealt with. Cheah *et al.* (1988) give the characteristics for the mixed palm oil mill effluent (POME) listed in Table 4.8. As shown in Table 1.2, this effluent is produced at the rate of 2.5 tons per ton of palm oil, and oil production by the two major producers is currently 6.2 million tons in Malaysia and 2.5 million tons in Indonesia. A number of biological treatment systems have been successfully implemented.

Table 4.8 Characteristics of palm oil mill effluent

Parameter	
Temperature	80—90°C
pH	4.0
BOD	25,000 mg/l
Suspended solids	19,000 mg/l
Total nitrogen	770 mg/l
Ammoniacal nitrogen	35 mg/l
Oil and grease	8,000 mg/l

Anaerobic and facultative ponds. Effluent passes successfully through the following elements (Chooi, 1985):

(1) A concrete de-oiling tank. Effluent at 60–70°C is held for 1.5 days. Some free oil and solids separate out, and the effluent composition is averaged.

(2) Two acidification ponds in series with two days retention in each. In the first pond liquor from the first anaerobic pond is added to supply seed bacteria, adjust pH and reduce the temperature.
(3) Two anaerobic ponds with holding times of 30 and 15 days where volatile fatty acids are converted to methane and carbon dioxide.
(4) Three shallow ponds for aerobic digestion with a total holding time of 16 days. BOD, nitrogen and suspended solids are reduced.

The final treated discharge has pH 8.7, BOD 100 mg/l, total solids 6000 mg/l and total nitrogen 85 mg/l.

Anaerobic tanks and aerated ponds. The process involves the following sequential elements:

(1) An acidification pond for one day.
(2) Anaerobic tank digestion for 20 days. Solids are periodically removed from the tank for land application as fertilizer.
(3) After one day in a settling tank the liqour is treated in a pond for 20 days with vigorous aeration by high-speed agitators.
(4) The liquor passes into a settling tank for two hours. Solids are periodically removed for land application, and the effluent is then fit for discharge.

Averaged analytical data for the final discharge from oil mills using this system over a prolonged period had the following ranges (in mg/l): BOD 56–99, suspended solids 214–537, oil and grease 75–154, total nitrogen 40–89 and ammoniacal nitrogen nil (Lim, 1985).

An economic analysis indicated that the fertilizer value of the sludge solids was greater than the operating costs of the treatment system. A flow diagram for the process described is given in Fig. 4.9.

Anaerobic tanks and land application. Anaerobic digestion tanks similar to those described by Lim (1985), were designed as a closed system in which the biogas (methane) generated could be utilized. A part of the gas is compressed and recirculated to the bottom of the tank to effectively mix the contents, as shown in Fig. 4.10 (Lim, 1985). The biogas is used to generate electricity. The digested effluent from the tanks may be used for land applications as a fertilizer without further treatment.

4.2.2.7 Biological treatment of olive oil mill effluent

A conventional aerobic digestion of olive oil mill effluent at 25°C for 7 days results in a final discharge of the following characteristics: pH 7.2, volatile acids 150 mg/l and ammoniacal nitrogen 20 mg/l. 83% of the organic substances were removed (Padilla, 1992).

The use of proprietary mixtures of bacteria for inoculating an aerobic treatment process has been described by Ranalli (1992).

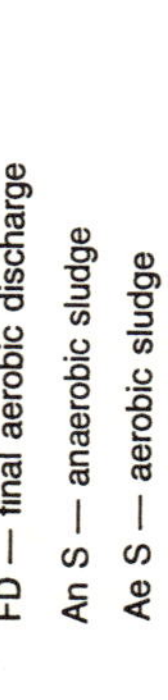

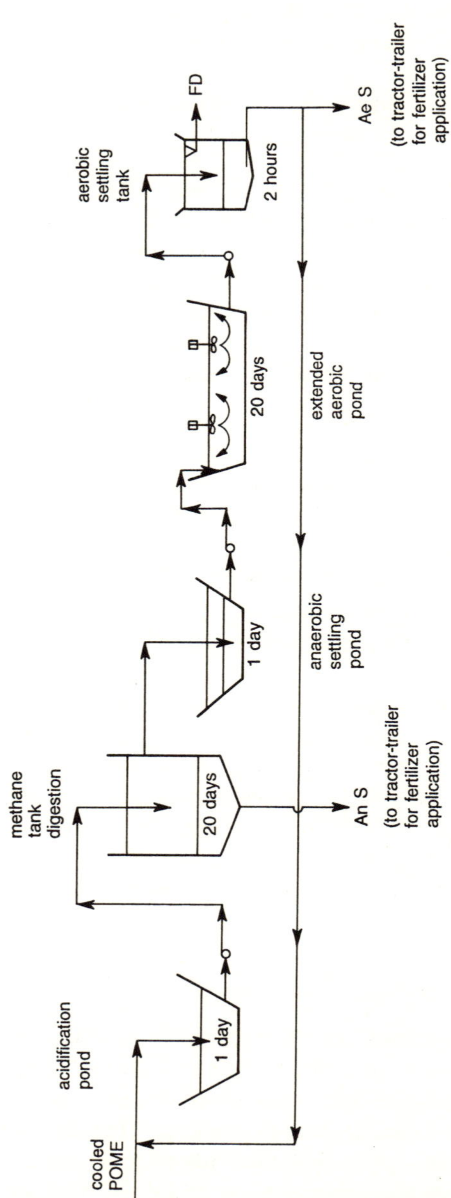

Fig. 4.9 Flow scheme of two-phase unstirred anaerobic contact and external aeration process.

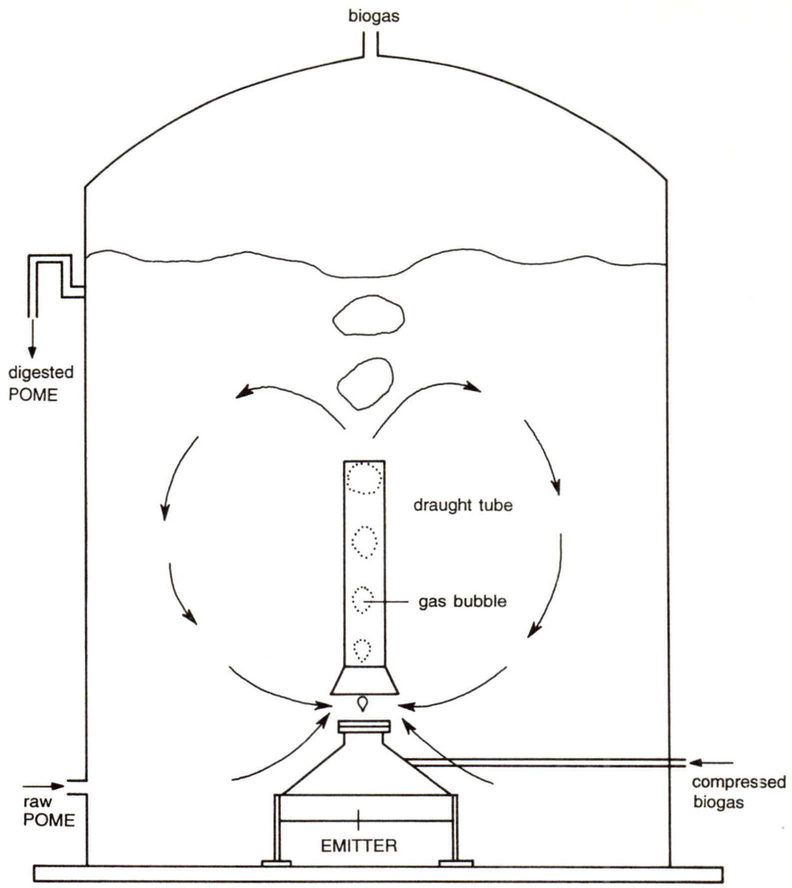

Fig. 4.10 Gas mixer showing flow pattern of digester content (POME—palm oil mill effluent).

Operating at pH 7 and 30°C, a reduction of up to 93% of the original COD was achieved with inoculates from ECOIL of Ravenna.

Alternatively an anaerobic digestion, similar to that described by Lepke (1988) (see Section 4.2.2.5 above) can be used. Fiestas Ros de Ursinos (1982) describes a process using a simple digestion tank, in which the charge is kept mixed by recirculating a part of the biogas generated. The process is operated at 35°C and pH 7.1–7.3. The effluent BOD of 31,000–44,000 ppm is reduced to between 3600 and 7600 ppm, on average by 80%. Methane generation is 0.85 m^3 per kg of BOD removed.

Difficulties are experienced in the anaerobic digestion of olive oil mill effluent because the phenolic substances present, derived from the fruit, inhibit the growth of the bacteria (Duran, 1991). A preliminary aerobic treatment with specific bacteria is recommended in order to remove most of the phenols before the anaerobic process. The initial level of phenols was 300 ppm (measured as caffeic acid); it was reduced by aerobic treatment to 103 ppm and after anaerobic treatment to 0.4 pp. Chemical oxygen demand was reduced from 50,000 ppm to 22,000 and 4000 ppm in the successive steps.

4.2.2.8 Treatment of wastes from oleochemical plants

The oleochemical industry produces wastes that pollute the atmosphere, water and dumping grounds. Liquid wastes can be formed in any branch of this industry. They can contain leakages and deposits from tanks, residues from pretreatment, condensate from vacuum fat splitting devices, water and condensate from fatty acids distillation, water condensate from concentration and distillation of glycerine, etc. The BOD is the most comprehensive criterion of the degree of pollution (Boyer, 1984); however, in the case of wastes from this industry the glycerine content must also be considered. In terms of the BOD 100 ppm of glycerine corresponds to 122 mg/dm^3 BOD. Other feature characteristics for such wastes are low pH, large content of dissolved organic compounds, fats and sulphates. The dissolved organic compounds are mainly glycerine, alcohols, glycols and surfactants. Soluble fatty acids are lethal for fish even at low concentrations. Heavy metals, used as catalysts for fat splitting or hydrogenation, are harmful even in trace amounts.

Counteracting these hazards comprises: thorough inspection of the equipment to avoid unnecessary losses, maximal recovery of fatty matter, utilization of devices for the removal of mist and droplets entrained during distillation, application of surface instead of barometric condensers, etc. Plants producing fatty acids and their nitrogen derivatives yield wastes of a very diversified composition. The wastes originate from the production of amines, quaternary ammonia salts, ethoxylated amines and fatty acids fractionation. They contain vegetable and animal fats. Purification of such wastes to a degree allowing their discharge to the municipal sewage system requires various methods. Even the thorough reduction of the amount of hexane solubles is difficult, since the wastes contain surfactants stabilizing the emulsions. As a result the separation of fatty matter by skimming is impossible. The method does not enable the fat content to be reduced below 100 mg/dm^3.

The manner of removal of the organic matter depends on whether it is dispersed or dissolved in water. Dispersed pollutants which were not eliminated in degreasers or centrifuges can be removed by neutralization and flocculation, introduction of polyelectrolytes and flotation. Special methods are used for the removal of dissolved substances, e.g. glycerine. These methods comprise e.g. oxidation with chlorine or with UV radiation.

The wastes from these plants usually pass through a stirrer averaging their composition and then through two devices, comprising three traps each, decanting and draining off the fats. Then the wastes are neutralized with an NaOH solution of the density 1.530 to pH 6.5–7.5 in apparatus equipped with air stirring. Wastes treated in such a way can have for example the following characteristics: pH 7.0, BOD 10,000 mg/dm^3, suspended matter content 300 mg/dm^3 and fat content 100 mg/dm^3. To discharge them to municipal wastes it is necessary to reduce their BOD down to 25 mg/dm^3 (Hamilton, 1969).

The processes of biodegradation by activated sludge of the following compounds have been examined: fatty acids, amines, nitriles, amino acids and dicarboxylic acids. The results proved that aliphatic C_1—C_{11} compounds are rapidly oxidized. The C_5—C_{10} acids revealed a certain toxicity. The C_{14}—C_{19} acids were initially toxic or neutral to all the sludges. The presence of unsaturated fatty acids significantly accelerated the oxidation. C_1—C_{10} fatty amines poisoned all the sludges. Amino acids behaved in various ways, but were usually easily oxidized (Malaney and Gerhold, 1970). Therefore biological treatment has been used for all the wastes except those from the production of quaternary ammonia salts, which are poisonous for the sludge. The wastes were first subjected to skimming and mechanical purification.

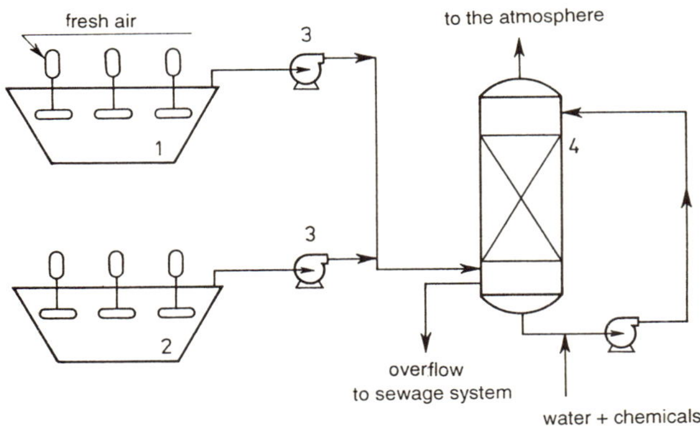

Fig. 4.11 Deodorization of fumes from biological treatment plant (Harp, 1975): 1—fresh air in through openings around aerator shafts, 2—aeration basins with solid metal covers each, 3—blower, 4—scrubber.

Two chambers aerated with 2 mg of oxygen per dm^3 at a 90% load are usually used for the biological treatment. The residence time of the waste in the chamber is 60 hours. The equipment allows the adjustment of pH and phosphorus content necessary for the growth of the microflora. Enrichment of the wastes with nitrogen is unnecessary, since the wastes contain ammonia. The outflow from the chambers is directed to a settling tank, where the separation of the sludge takes place. The sludge is recycled to the aerated chamber. The outflow from the settling tank is discharged to the municipal sewage system. The content of hexane soluble substances is then equal to ca. 102 mg/l. The entire installation produces noxious odours, therefore it is roofed, and the fumes that accumulate inside are neutralized in a way illustrated in Fig. 4.11.

Adsorption on active carbon. Adsorption on active carbon can be the next step after the biological treatment. Columns packed with active carbon or fluidized beds are used for this purpose. The flow rate is usually 203.5–325 $dm^3/min \cdot m^2$. It has been established that the maximal adsorption is equal to 0.5 kg COD/kg. In the case of detergents (alkylbenzenesulphonates) the adsorption can be equal to 0.03 kg/kg of carbon.

4.2.2.9 Profitability of waste treatment

The calculation of the profitability of waste treatment is limited to a comparison of investment costs, repayment of credit and of operational costs, with the costs of discharging the wastes to a municipal sewage system with possible charges for excessive pollution and sludge removal (Brauch, 1982).

The agencies charge the industry on the basis of the following indicators of the extent of waste pollution: BOD, COD and suspended matter content. The charge depends also on the purity class of the receiver (valid in some countries). The total annual charge should be compared to the costs of waste treatment by the producers. The commercial value of the recovered fat or other wastes is also significant. A waste treatment plant can be financially self-sufficient, and sometimes it can even make profits.

In large oil plants, utilizing flocculation and modern centrifuges for the concentration and refining of the recovered fatty matter, the value of the regenerated fats covers 60% of the costs of waste treatment and regeneration of the fats. For instance, fat recovered from a large vegetable oil refinery in the USA has the following characteristics (Seng and Kreuzer, 1975):

water	0.3%
ether soluble substances	98.3%
ether insoluble substances	0.3%
ash	0.13%
FFA	21.9% of the fatty matter
UM	2.5% of the fatty matter
FA *c.p.*	34.9°C
SV	198.2

The costs of waste treatment comprise 38% for chemicals and 46% for labour costs, while the expenditure for fat recovery comprises: 29% for sludge removal, 44% for labour costs.

Fatty wastes from meat processing plants are treated in a single operation, the fatty matter content being reduced from 3,000–5,000 mg/kg to less than 100 mg/kg. The recovered fatty matter is passed through a concentrating settling tank, where the concentration of the hexane-soluble substances increases from 10 to 45%. The value of the recovered fat is greater than the expenditures.

The investment costs are lower when the installation has a high efficiency. This is due to a smaller area of settling tanks. Flotation allows a much better utilization of the area than sedimentation devices. Although the cost of flotation equipment is higher than that of sedimentation devices, the concentration of the sludge obtained is higher, hence the costs of further processing are lower. The expenditure for energy spent on lifting the wastes is higher in flotation devices. On the other hand, flotation assures a much better efficiency factor, both with respect to the degree of purification, and to the concentration of the sludge.

The costs of waste treatment depend to a large extent on the total amount of water consumed, hence it is very significant to save it. According to Seng (1980), rational water management allows saving 50% of water.

4.2.2.10 Standards of wastes purity

The composition of wastes discharged to municipal sewage systems must meet the requirements of the respective regulations. The permitted concentration of vapours and gases that can be released from the wastes to the air contained in the sewage system is also limited. In the case of petrol it is 0.2 mg/dm^3 of air.

The requirements concerning the purity of refinery wastes discharged to municipal sewage systems differ in various countries. For example, in the United States they are the following:

	mg/dm^3
BOD	< 200
suspended matter	< 200
fatty matter	< 100
pH	6.5–9
temperature	65°C

Refinery wastes (without sanitary wastes) are usually treated together, except for the wastes from soapstock acidification and the deodorizer scum.

4.3 SOLID WASTES

Solid wastes are mainly the spent bleaching earth, sludges, glycerine and fatty acid distillation residue, and sometimes remnants of damaged packages. They are less dangerous for the environment, because usually they do not contain toxic components and therefore can be disposed of in the same way as municipal solid wastes. Their neutralization consists in appropriate treatment or storage so that noxious substances cannot come into contact with the human environment. The neutralization may consist of converting the wastes into a harmless form, or into a useful form after mixing with other materials.

Bleaching earth. In a typical case the mass of earth used is 1% of the oil processed and the earth contains up to 28% of oil. In such a case a refinery processing 10 t/d produces daily 3336 kg oiled bleaching earth cakes, containing 936 kg oil. The energy value of this oil is 352 MJ. The remaining components are: 10% fat, 65% silicates and other mineral components. For transport purposes the earth should contain ca. 25% water. However, the transport of bleaching earth in the suspended form is difficult. It has been proposed therefore to pyrolyse the oiled cakes in closed autoclaves and to burn the released gases in boilers. Removal of the majority of the fat from oiled bleaching earth yields a waste that can be discharged without any limitations at municipal dumping grounds, since the residue of fat does not create a danger of self-ignition. A product of a composition similar to bentonite is obtained as a result of boiling the earth with soda. Its suspension in water is completely insoluble and is difficult to dehydrate by filtration or centrifuging. These properties are changed after decreasing the pH to 4 and adding calcium chloride. The water content is thus reduced, so that the consistency of the sludge is changed to semi-fluid and water permeable. Such a waste can be used for cement production, for obtaining SiO_2 and as a treatment for sandy soils. The results obtained with maize growing on sandy soils treated with bentonite sludge and with the suspension with calcium sulphate added are promising. The addition can be equal to 100 kg of the bentonite suspension and ca. 200 kg of the calcium sulphate treated sludge per 1 m^3 of the sandy soil (Ong, 1983).

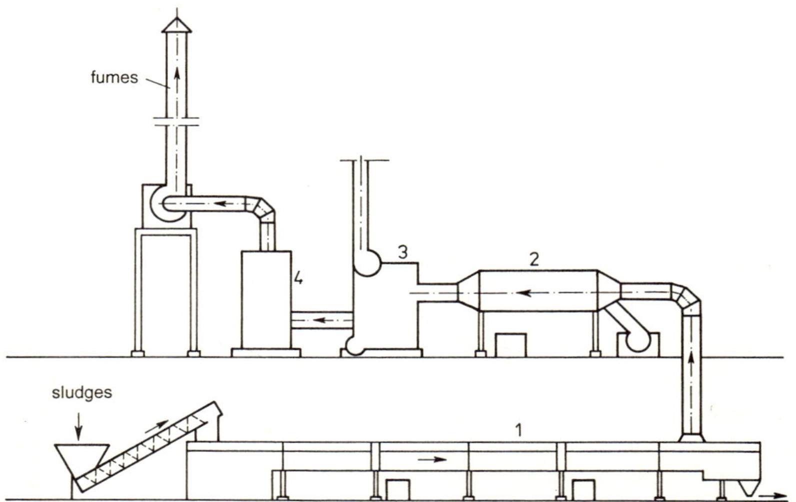

Fig. 4.12 Sludge incineration (Geldner, 1983): 1—reactor heated by IR radiation, 2—secondary incineration furnace, 3—heat regenerator, 4—scrubber.

Sludges. They are noxious wastes from oil refineries. Their origin can be various:

—from sediments removed periodically,
—from neutralization — as bottom sludges and sometimes dispersed sludges,
—scums from flotation of acid water.

Crude sludges contain 3–10% d.m. on average. They are processed into a dry form in various ways: by drying in air, mixing with spent earth, screening or pressing, and can then be discharged at dumping grounds. Taking into account the costs of these operations it seems that the best solution is filtration at ca. 40°C with an addition of 10% of deoiled spent bleaching earth. The filtration cakes obtained contain ca. 35% water.

Sludges of large volume can be incinerated. One of the methods of heating consists in utilization of the IR radiation. Incineration proceeds continuously. Utilization of the heat produced significantly increases the profitability. The volume of ash leaving the apparatus is much smaller than that of the sludge from which it was formed (Geldner, 1983) (Fig. 4.12).

Solid wastes are sometimes incinerated in a fluidized bed. Except for the catalyst they can be incinerated provided that their energetic value exceeds 8.2 kJ/kg. The residue after incineration can be used as a filling material. The gases released during incineration should be discharged to the atmosphere after passing through a dust separator and wet fume catcher.

REFERENCES

Axberg C. *et al.* (1980), *Prog. Wat. Tech.*, **12**, 371.
Beal R. *et al.* (1973), *J. Am. Oil Chem. Soc.*, **50**, 260.
Becker K. (1972), *J. Am. Oil Chem. Soc.*, **49**, 40A.
Boyer M. (1980), In: *Handbook of Soy Oil Processing and Utilization*, Erikson D. *et al.* (Eds.), Am. Soybean Ass., St Louis and Am. Oil Chem. Soc., Champaign, 531.
Boyer M. and Fithian T. (1980), In: *Handbook of Soy Oil Processing and Utilization*, Erikson D. *et al.* (Eds.), Am. Soybean Ass., St Louis and Am. Oil Chem. Soc., Champaign, 532.
Boyer M. (1984), *J. Am. Oil Chem. Soc.*, **61**, 297.
Brauch V. (1982), *Fette, Seifen, Anstrichm.*, **84**, 547.
Bystram K. (1976), *Engineer's Handbook. Fat Industry* (in Polish), WNT, Warszawa.
Bystram K. (1979), *Biul. Inform. Inst. Przem. Mięsnego i Tłuszczowego,* **17** (2) 45.
Cantrell M. and Keller H. (1975), *J. Am. Oil Chem. Soc.*, **52**, 13A.
Cheah S. C., Ma A. N., Ooi L. C. L. and Ong A. S. H. (1988), *Fat Sci. Technol.*, **90**, 536.
Choffel G. (1976), *J. Am. Oil Chem. Soc.*, **53**, 446.
Chooi C. F. (1985), *Proc. Rev. Palm Oil Mill Effluent Technology,* Kuala Lumpur, 1984, Palm Oil Research Institute of Malaysia, p. 53.
Conze E. (1982), *Fette, Seifen, Anstrichm.*, **84**, 318.
Crauer L. (1970), *J. Am. Oil Chem. Soc.*, **47**, 210A.
Daković S. *et al.* (1985), *Fette, Seifen, Anstrichm.*, **87**, 11.
Duran R. M., Padilla R. B., Martin A. M., Fiestas Ros de Ursinos J. A. and Mendoza J. A. (1991), *Grasas y Aceites,* **42**, 271.
Eckenfelder W. *et al.* (1970), *Waste Treatment*, Vol. 3, Texas University, Austin, p. 4.
Eisenhauer R., Beal R. and Griffin E. (1970), *J. Am. Oil Chem. Soc.*, **47**, 137.
Farr W. and Bolhofner K. (1980), *J. Am. Oil Chem. Soc.*, **57**, 6.

Fiestas Ros de Ursinos J. A., Gamero R. N., Cabello R. L., Buendia A. J. G. and de Jauregui G. M. M. S. (1982), *Grasas y Aceites,* **33**, 265.
Geldner G. (1983), *Fette, Seifen, Anstrichm.*, **85**, 556.
Goodrich W. (1980), In: *Handbook of Soy Oil Processing and Utilization*, Erickson D. *et al.* (Eds.), Am. Soybean Ass., St Louis and Am. Oil Chem. Soc., Champaign, 521.
Hamilton H. (1969), *J. Am. Oil Chem. Soc.*, **46**, 334.
Harp E. (1975), *J. Am. Oil Chem. Soc.*, **52**, 4A.
Holt E. and Brown A. (1971), *J. Am. Oil Chem. Soc.*, **48**, 4A.
Imhoff K. (1982), *Municipal Sewage System and Waste Treatment. A Handbook* (in Polish), Arkady, Warszawa.
Kenyon W. (1970), *J. Am. Oil Chem. Soc.*, **47**, 402A.
Koziorowski B. (1980), *Industrial Liquid Waste Treatment* (in Polish), WNT, Warszawa.
Krause A. (1978), *Fette, Seifen, Anstrichm.*, **80**, 77.
Krause A. (1982), *Fette, Seifen, Anstrichm.*, **84**, 536.
Kroll S. (1980), *Fette, Seifen, Anstrichm.*, **82**, 357.
Kubicki M. (1965), *Tłuszcze i środki piorące*, **2**, 289.
Lacy W. *et al.* (1975), *J. Am. Oil Chem. Soc.*, **52**, 20A.
Lau J. and Neelsen W. (1980), *Fette, Seifen, Anstrichm.*, **82**, 166.
Lepke P. (1988), *Fat Sci. Technol.*, **90**, 576.
Liebe H. G. and Munch E. W. (1986), *Fat Sci. Technol.*, **88**, 537.
Lim K. H. (1985), *Proc. Rev. Palm Oil Mill Effluent Technology,* Kuala Lumpur, 1984, Palm Oil Research Institute of Malaysia, p. 11.
Lim K. H., Quah S. K., Gillies D. and Wood B. J. (1985), *Proc. Rev. Palm Oil Mill Effluent Technology,* Kuala Lumpur, 1984, Palm Oil Research Institute of Malaysia, p. 42.
Lindh L. and Dahlen J. (1989), *J. Am. Oil Chem. Soc.*, **66**, 972.
Lüde R. (1957), *Die Raffination von Fetten und Fetten Öle*, Steinlopff, Dresden.
Malaney G. and Gerhold R. (1970), *J. Water Pollut. Control Fed. Res.*, Suppl., **41** (2, part 2).
McDermott G. (1976), *J. Am. Oil Chem. Soc.*, **53**, 449.
McDermott G. (1976), *J. Am. Oil Chem. Soc.*, **56**, 789A.
Montenegro H. M. (1985), *J. Am. Oil Chem. Soc.*, **62**, 259.
Neuner W. and Holt E. (1975), *J. Am. Oil Chem. Soc.*, **52**, 17A.
Ong J. (1983), *J. Am. Oil Chem. Soc.*, **60**, 314.
Padilla R. B., Barrantes D. and Gonzales L. (1992), *Grasas y Aceites,* **43**, 20.
Pfeiffer H. (1982), *Fette, Seifen, Anstrichm.*, **84**, 556.
Radandt S. (1982), *Fette, Seifen, Anstrichm.*, **84**, 235.
Ranalli A. (1992), *Grasas y Aceites,* **43**, 16.
Rautenbach R. *et al.* (1984), *Fette, Seifen, Anstrichm.*, **84**, 18.
Richter H. (1981), *Fette, Seifen, Anstrichm.*, **83**, 10.
Schmidt-Holthausen H. (1979), *Fette, Seifen, Anstrichm.*, **81**, 253.
Schöning H. and Neumann U. (1983), *Fette, Seifen, Anstrichm.*, **85**, 551.
Seng W. (1980), *J. Am. Oil Chem. Soc.*, **57**, 926A.
Seng W. and Kreuzer G. (1975), *J. Am. Oil Chem. Soc.*, **52**, 9A.
Skalski K. (1975), *Przemysł Spożywczy,* **29**, 196.
Steinhauer R. and Graalmann M. (1981), *Fette, Seifen, Anstrichm.*, **83**, 364.
Swern D. (1964), *Bailey's Industrial Oil and Fat Products*, Interscience Publ., New York.
Team Work (1969), *Water and Wastes in Food Industry* (in Polish), Wyd. Przem. Lekkiego i Spożywczego, Warszawa.
Wall W. (1982), *Oil Recovery Techniques*, Envirex, Waukesha.
Zimińska H. *et al.* (1984), Food Preservation and Technical Microbiology Division, Technical University of Gdańsk (in Polish).
Zockoll C. (1979), *Fette, Seifen, Anstrichm.*, **81**, 268.
Zockoll C. (1982), *Fette, Seifen, Anstrichm.*, **84**, 558.

Index

acid hydration 39–40
acid oil 54
acid value 37, 51, 58, 75, 87, 112, 122, 140, 142, 143, 147, 148, 155, 158, 159, 160, 161, 162, 163, 167, 174
activated sludge 201–204
alcohols 109, 115, 158, 210
alkali refining (alkali neutralization) 54, 66, 67, 149
alkali refining losses 66, 67
alkanolamides 116, 117
aluminium soaps 153
ammonia soaps 156
anaerobic treatment 204
animal fats 36, 68, 72, 89, 120, 122, 126, 150, 151, 155, 156, 159, 164
animal fats, refining 74–76
animal fats, rendering 122, 123
anisidine value (AnV) 37
Aspergillus flavus 134
autoclave oil 148

back fat 121
beef tallow 72, 80, 121
biodegradation 51, 205, 206, 211
biogas from POME 207, 209
biological ponds 201, 204
bleaching 55
bleaching earth 56
bleaching earth, oil from 147
bleaching earth, regeneration 59
bleaching earth, spent 170
bleaching earth, treatment 214
bleaching earth, utilization 147, 148
BOD 51, 53, 174, 178, 182–187, 189–192, 194, 198, 200, 202–210, 212, 213
bone fat composition 162
bone fats 13, 84, 125, 161, 162

calcium soaps 149
Candida rugosa 83
carbon dioxide, supercritical 27–29
castor oil 83, 87, 102, 105–106, 155, 156, 174
castor oil, dehydration 106
catalyst, spent 169
centrifuging 48
coagulation 186, 187, 198, 199
cocoa butter 57, 140
coconut meal 180
coconut oil 56, 60, 64, 72, 80, 92
COD 66, 174, 184, 186, 198, 199, 205, 212
condensers, deodorizer 65, 66
corn 36
cotton seed 36, 134, 143
cotton seed lecithin 143
cotton seed meal 134
cotton seed oil 93, 94, 148
countercurrent extraction 58
crude oil 20, 39, 40, 44, 46, 47, 48, 49, 51, 52, 55, 58, 66, 67, 70, 107, 144, 148, 149, 170, 189

decarboxylation 85
degras 161, 162, 163
degumming 35–39
degumming of waste fats 76
dehulling, soyabeans 15
dehulling, using fluidized beds 16, 17
deodorization 59
deodorizer condensate 60
deodorizer condensers 65, 66
deodorizer distillate from palm oil 149
deodorizer scum 61–63
desludging by acid hydrolysis 40
desludging by bleaching earth 61
desolventizer 23
desolventizing 27
detergent fractionation fatty acids 95, 96
dewaxing of sun flower oil 48–50
DFA (distilled FA) 73, 87, 96, 97, 98, 159
dimers 118, 119
dimers of esters 118, 119
dimers of fatty acids 118, 119

distillation forerunnings 157
distillation residue, composition 154, 155
distillation residue, utilization 154, 155
dust 78
dust explosions 178

edible oil 13
emulsifier, from bleaching earth 170
enzymatic hydrolysis 83
epoxidized oils 144
erucic acid 40, 142, 148, 175
even carbon number FA 157, 158
explosion limits 178–180
explosions, solvent-air 179, 180
extraction 25, 27, 28, 29, 30, 37, 38, 46, 58
extraction meal 15, 23
extraction, supercritical fluid 27–29
extraction, using enzymes 30

fat amides 117
fat, from furred animals 126, 164
fat, from sewers 126–128
fat splitters, enzyme 83, 84
fat spitters, high pressure 81–83
fats, pressing 19, 29, 31
fats, refining 29, 73–77, 96, 146, 180–181, 183
fats, saponification 102
fats, splitting (hydrolysis) 79–84, 150
fatty acid methyl esters 100, 111, 112
fatty acid monoglycerides 112–115
fatty acids (FA) 52, 54, 62, 63, 65–68, 70, 79, 80, 84–99, 100, 102, 106, 107, 108, 109, 111, 112, 113, 115, 116, 117, 118, 119, 124, 128, 139, 140, 144, 145, 146, 148, 149, 151, 153, 154, 155, 156, 157, 160, 162, 163, 164, 174, 175, 177, 181, 182, 183, 186, 210
fatty acids, esterification 111–113
fatty acids, ethoxides 115
fatty acids, fractionation by crystallization 93–96
fatty acids, fractionation by distillation 88–93
fatty acids, hydrogenation 96–99
fatty acids, saponification 102
fatty acids, volatile 62, 63
fatty alcohols 99–101
fatty alcohols, distillation 101
fatty alcohols, obtaining 99–101
fatty amines 117, 118
fatty nitriles 117
feed meals, composition 135–138
filtration 47, 49, 78, 187, 195
fish oil 13, 36, 48, 57, 89, 123, 124, 160, 161
fish processing wastes 173
flocculation 186, 198, 199, 206, 210, 212
flotation 186, 198, 199, 200, 201, 206, 210, 213, 215
forerunnings 157
fractional distillation 89, 90, 91, 92
free fatty acids (FFA) 19, 28, 40, 44, 45, 46, 51, 52, 56, 59, 60, 66, 67, 68, 70, 73, 75, 79, 80, 81, 139, 145, 146, 148, 149, 160, 164, 170, 171, 173, 174, 175, 190, 201, 212
frying oil 170
fumes 177, 179, 180–182

glucosinolates, repeseed 135
glycerine 79, 80, 81, 83, 100, 102, 111, 115, 116, 117, 150–154, 210, 213
glycerolysis 112, 113
gums 139, 143, 170

halogenation 46, 141
high erucic rapeseed 19, 92, 142
hulls 168–169
hydratable phospholipids 36, 144
hydration 36, 37, 38, 40, 44, 56, 67, 76, 78, 139, 140, 141, 189
hydrogenation 70, 71
hydrogenation catalyst-spent 169
hydrolysis (see also fat splitting) 29, 46, 79–84
hydroxylation 46

industry fumes 177
iodine value (IV) 57, 71, 87, 91, 92, 94, 96, 112, 122, 124, 142, 145, 147, 149, 155, 157, 158, 159, 160, 161, 162, 163

jojoba seed 29

lanolin 103, 104
lanolin soaps 158, 159
lard 120
lauric oil 145
leaf fat 120, 121
lecithin 11, 13, 29, 35–40, 55, 138–143
lecithin, composition and properties 138–142
lecithin, obtaining 43–46

lecithin, refining 46
lecithin, utilization 142–144
linseed 30
linseed oil 42, 48, 57, 72, 87
liquid wastes 170, 171, 182–213
liquid wastes, coagulation 186, 187, 198–199
liquid wastes, fat separation 189
liquid wastes, filtration 187, 195
liquid wastes, flocculation 198–199
liquid wastes, flotation 199–201
liquid wastes, neutralization 186, 192
liquid wastes, refining 183, 184
liquid wastes, sedimentation 186, 187
liver oils 124
lye glycerine 153

magnesium soaps 149
maize 146, 167, 170, 171, 214
maize oil 36, 48, 68, 144
margarine 13, 71, 143, 190
marine fats 124
meal 13, 18, 20, 21, 22, 23, 24, 25, 26, 29, 38, 39, 40, 134–138, 168, 171, 179, 180, 183
meat processing wastes 174, 175
meat tissue fats 120
mink fat 75
monomers 158

natural oil 112, 146
neutralization, ammonia 55
neutralization in miscella 70
neutral fat 80, 152
neutral oil 66, 67, 70, 119, 144, 145, 148, 190
non-edible fats 71, 72, 73, 74, 79, 99, 102, 111, 122, 125, 127, 150, 159
non-hydratable phospholipids 37, 38, 144

odd carbon number FA 158
oil cake 13, 18, 19, 26, 30
oils, bleaching 55–58
oils, degumming 35–43, 45, 52, 55
oils, deodorization 59–66, 67, 68, 148, 189
oils, desludging 36, 40
oils, epoxidized 120
oils, hydrogenation 70, 71, 72
oils, polymerized 119
oils, refining 46–70
oils, superdegumming 42–43
oils, winterizing 70
oleochemicals, wastes 172
olive 36
olives, processing 34, 35

palm fruit, processing 31–34
palm kernel oil 55, 56, 63
palm oil 36, 60, 64, 67, 68, 72, 145, 147, 148, 149
palm oil, physical refining 149
peanut oil 36, 42, 80
peroxide value (PV) 37, 43, 58, 75, 157, 159, 160, 162, 163, 167, 171
pesticides 56, 60, 175
phospholipids 29, 35–39, 41, 42, 43, 45, 46, 56, 66, 67, 68, 79, 139, 143, 191
physical neutralization 67–70
physical refining 43, 56, 68, 69, 70, 149, 175
polyaromatic hydrocarbons 55, 56
polymerized oils 119
ponds, biological 201, 202, 204
post-hydration gums 138, 144
post-refining acids 11, 13, 83, 120, 134, 137, 146, 175
post-refining FA 54, 150, 161
potassium soaps 152
poultry fats 120, 126
pre-pressing 20, 21
pressed oil 19, 45, 46, 48
pressed oil, degumming 39
pressing 19

rapeseed 19, 21, 24, 26, 30, 36, 39, 43, 92, 93, 134, 138, 142, 167, 168, 179, 183
rapeseed hull 18, 168
rapeseed lecithin 44, 141–143
rapeseed meal 21, 23, 136–138
rapeseed meal, composition 136
rapeseed oil 19, 39–42, 53, 60, 67, 68, 142, 144, 148, 150, 153
rapeseed oil, degumming 39
rapeseed oil, desludging 40
rapeseed oil, high erucic 59, 94, 148
rapeseed oil, low erucic 68, 158
refining coefficient 66
residual solvent 24, 25
resin acids 107
rice oil 36
ricinus 84
ricinus oil 87

salad oil 48
saponification value (SV) 79, 127, 155, 158, 159, 160, 161, 162, 163, 212

saturated 84, 86, 158, 159, 161
scrubber, vapours 63, 196
sedimentation 46, 187, 189
sediments, tank 170, 171
seed impurities 167, 168
seed pretreatment 13–18
sewer grease 75, 126–128
SFA (FA from splitting) 79, 83, 84, 87, 96, 98, 151, 154
slaughter fats 159–160
sludge, activated 203, 204
sludges 11, 122, 172, 190, 200, 201, 202, 211–215
soap lyes 102, 152–153
soaps 50, 102, 106, 116, 145, 150, 152, 153, 159, 190, 193, 197
soapstock 50, 51–53, 57, 60, 67, 70, 75, 96, 144–146, 152, 180, 183, 185, 190, 192, 197, 213
soapstock, composition and properties 144–145
soapstock, hydrolysis 53
soapstock, utilization 145, 146
sodium soaps 152, 158, 159
solid wastes 213–215
solvent extracted oil 39
solvent extraction 20, 21, 22, 24, 27, 29, 30, 45, 49, 183, 184, 190
solvent losses 22
solvent recovery 26
solvent, residual 24, 25, 180
soyabean 15, 17, 18, 20, 21, 24, 26, 36, 37, 43, 45, 60, 134, 137, 138, 146, 178, 184, 185
soyabean lecithin 43–46, 138, 139, 140, 142, 143
soyabean meal 136, 137, 138, 180
soyabean oil 36–38, 40, 42, 58, 60, 67, 72, 120, 138, 140, 144, 145, 148, 190
soyabean oil, degumming 36–39
sprinkled bed 205–206
suint (see also wool wax) 158, 159, 164
suint, saponification 104
sulphate soaps 107, 150
sulphation 115, 116
sulphonation 115, 116
sunflower 15, 18, 19, 30, 36, 137, 168
sunflower lecithin 143
sunflower meal 136, 137, 147, 180
sunflower oil 36, 42, 56, 64, 68, 70, 170
sunflower oil, dewaxing 48–50
sunflower seed hull 18, 168–169
superdegumming 42, 43
surfactants 77, 79, 95, 100, 115, 116, 194, 198, 210
sweet waters 150–152

tall oil 71, 102, 107–109, 158, 162–163
tall oil, distillation 108, 110
tall oil, fractionation 109
tall oil pitch 155, 156
tallow 75, 83, 120, 121, 146, 155, 162
tannery fats 13
tanning wastes 124
technical oil 168
tung oil 106

unsaturated FA 45, 77, 86, 90, 118, 155, 158, 159, 211
utilization grease 72, 75, 84, 122, 123, 161, 162

waste fats, refining 75–79
waste treatment 185–215
waste treatment, biological 196, 201–206
waste treatment, chemical 196–197
waste treatment, mechanical 194–196, 202
waste treatment, profitability 212
wastes, neutralization 186, 187, 192
wastes, refineries 183, 184
wax 48–50, 70, 107, 201
wool wax (see also suint) 102–104

ELLIS HORWOOD SERIES IN
APPLIED SCIENCE AND INDUSTRIAL TECHNOLOGY

Series Editor: Dr D. H. SHARP, OBE, former General Secretary, Society of Chemical Industry; formerly General Secretary, Institution of Chemical Engineers; and former Technical Director, Confederation of British Industry.

series continued from front of book

PROFIT QUALITY: The Essentials of Industrial Survival
P.W. MOIR, Consultant, West Sussex
EFFICIENT BEYOND IMAGINING: CIM and its Application for Today's Industry
P.W. MOIR, Consultant, West Sussex
COLOUR DYNAMICS AS A SCIENCE
ANTAL NEMCSICS, Professor at the Technical University of Budapest, Hungary
BY-PRODUCTS AND WASTE MATERIALS IN FAT TECHNOLOGY
HENRYK NIEWIADOMSKI (deceased), Technical University of Gdansk, Poland, and
HANNA SZCZEPANSKA (deceased), formerly of the Institute of Industrial Chemistry, Warsaw, Poland
TRANSIENT SIMULATION METHODS FOR GAS NETWORKS
A.J. OSIADACZ, UMIST, Manchester
THE MECHANICS OF WOOL STRUCTURES
R. POSTLE, G.A. CARNABY and S. de JONG, Department of Textile Technology, University of New South Wales, Kensington, Australia
MICROCOMPUTERS IN THE PROCESS INDUSTRY
E.R. ROBINSON, Head of Chemical Engineering, North East London Polytechnic
BIOPROTEIN MANUFACTURE: A Critical Assessment
D.H. SHARP, OBE, former General Secretary, Society of Chemical Industry; formerly General Secretary, Institution of Chemical Engineers; and former Technical Director, Confederation of British Industry
QUALITY ASSURANCE: The Route to Efficiency and Competiveness, Second Edition
L. STEBBING, Quality Management Consultant
QUALITY MANAGEMENT IN THE SERVICE INDUSTRY
L. STEBBING, Quality Management Consultant
INDUSTRIAL CHEMISTRY
E. STOCCHI, Milan, with additions by K.A.K. LOTT and E.L. SHORT, Brunel
REFRACTORIES TECHNOLOGY: Manufacture and Industrial Application
C. STOREY, Refractories Consultant
COATINGS AND SURFACE TREATMENT FOR CORROSION AND WEAR RESISTANCE
K.N. STRAFFORD and P.K. DATTA, School of Material Engineering, Newcastle upon Tyne Polytechnic, and C.G. GOOGAN, Global Corrosion Consultants Limited, Telford
TEXTILE OBJECTIVE MEASURMENT AND AUTOMATION IN GARMENT MANUFACTURE
G. STYLIOS, Department of Industrial Technology, University of Bradford
PREPARATIVE AND PROCESS-SCALE LIQUID CHROMATOGRAPHY
G. SUBRAMANIAN, Department of Chemical Engineering, Loughborough University of Technology
INDUSTRIAL PAINT FINISHING TECHNIQUES AND PROCESSES
G.F. TANK, Educational Services, Graco Robotics Inc., Michigan, USA
MODERN BATTERY TECHNOLOGY
Editor: C.D.S. TUCK, Alcan International Ltd, Oxon
FIRE AND EXPLOSION PROTECTION: A Systems Approach
D. TUHTAR, Institute of Fire and Explosion Protection, Yugoslavia
COSMETICS AND TOILETRIES: Development, Production and Use
Editor: WILFRIED UMBACH, Director, Chemical Research, Henkel KGaA, Dusseldorf, Germany
PERFUMERY TECHNOLOGY 2nd Edition
F.V. WELLS, Consultant Perfumer and former Editor of *Soap, Perfumery and Cosmetics*, and M. BILLOT, former Chief Perfumer to Houbigant-Cheramy, Paris, Président d'Honneur de la Société Technique des Parfumeurs de la France
THE MANUFACTURE OF SOAPS, OTHER DETERGENTS AND GLYCERINE
E. WOOLLATT, Consultant, formerly Unilever plc
HOW TO TURN ROUND A MANUFACTURING COMPANY
BRIAN HALFORD WALLEY, Former Finance and Site Director, Ferodo Ltd